Advance praise for
The Once and Future World

"A gripping and convincing look at the nature that humans lost and the perspective that we gained. MacKinnon leaves us wanting to be wilder." Jennifer Jacquet, author of the *Guilty Planet* blog at *Scientific American*, assistant professor of environmental studies at NYU

"MacKinnon is an eloquent guide through landscapes wild and tame. He takes the reader backwards through evolutionary time and forward into a delicate and unknown future. I devoured this book in a day and closed its covers marveling at our planet's incredible abundance. Natural history at its best." Charlotte Gill, author of *Eating Dirt*

"Henry David Thoreau warned us, in 1862, that not in *wilderness* but in *wildness* is the preservation of the world. There's a difference. In *The Once and Future World*, J.B. MacKinnon brings this distinction up to date. Wilderness may be gone forever, but wildness can be recovered, and it is time to get to work." George Dyson

"A lean, elegant and powerful essay on what we have done to the world—and what we might do to set things right. J.B. MacKinnon has made me think in new ways about our self-destructive trashing of the 'luckless garden' into which we were so lucky to be born." Ronald Wright, author of *A Short History of Progress*

"A re-enchantment with the natural world may be a necessary prerequisite to the changes we must make to keep that natural world more or less intact. This is deep and lovely thinking and writing." Bill McKibben, author of *Oil and Honey: The Education of an Unlikely Activist*

"*The 100-Mile Diet* forever changed the way I see a plate of food and it is still with me today. *The Once and Future World* changed the way I see *everything*. One can only hope it spawns a movement like *The 100-Mile Diet* did—a moment of re-imagining, re-wilding and coming home." Leanne Allison, filmmaker (*Being Caribou, Finding Farley, Bear 71*)

"This book should make your blood run cold; or boil with furious rage against the despoilers of our planet. But perhaps all is not yet lost. MacKinnon tells us that the crisis in the natural world is not yet fatal . . . but it's waiting. And then he tells us most convincingly what we can and must do to stop the rot. This is a handbook for those who hope to see the earth, and man, remain alive together." Farley Mowat

"Like Peter Matthiessen, Barry Lopez and Tim Flannery, J.B. MacKinnon is an exceptional writer with an intense passion for the natural world. In *The Once and Future World*, MacKinnon combines eloquent storytelling with painstaking research to provide a persuasive argument for the need to not only protect the wildness we have today, but to restore at least some of the abundance we have lost. It may be too late to bring back the Tasmanian tiger, but, as MacKinnon writes, there's still time to create a planet that is far richer in natural wonders." James Little, former editor of *Explore* magazine

"J.B. MacKinnon is one of the finest essayists of the natural world writing today." Andrew D. Blechman, managing editor of *Orion* and author of *Pigeons: The Fascinating Saga of the World's Most Revered and Reviled Bird*

"This book is a delight. MacKinnon shows us afresh the world we thought we knew through a kaleidoscopic lens of startling facts, illuminating insight and flat-out wonderful writing." John Vaillant, author of *The Golden Spruce* and *The Tiger*

Praise for *The 100-Mile Diet*
(co-authored with Alisa Smith)

NATIONAL BESTSELLER

Winner of the Roderick Haig-Brown Regional Prize
Finalist for the Hubert Evans Non-Fiction Prize

"A compelling, relevant story without preaching or darkening our minds with guilt." *The Vancouver Sun*

"Eating locally isn't just a fad like the various diets advertised on late-night TV—it may be one of the most important ways we save ourselves and the planet." Dr. David Suzuki

"Smith and MacKinnon are gifted writers, and their inexperience at food sourcing makes them naturally more sympathetic." *Winnipeg Free Press*

"Engaging, thoughtful essays packed with natural, historical and personal detail." *The New York Times*

"Good writing trumps anything, and in Alisa Smith and J.B. MacKinnon's *100-Mile Diet* it brushes aside the inescapable but faint touch of preachiness. It's a finely rendered account of a year of eating locally." *Maclean's*

"*The 100-Mile Diet* is inspiring in its honest striving to discover what has been all but lost." *The Gazette*

"A fascinating personal journey narrated by two excellent writers that take us for an informative and highly entertaining trip into a world that just a generation ago was commonplace but now seems exotic." *Edmonton Journal*

Praise for *Dead Man in Paradise*

Winner of the Charles Taylor Prize for Literary Non-Fiction

"*Dead Man in Paradise* . . . works as travelogue, thriller and much-needed antidote to the ways in which history is often buried and forgotten." *Quill & Quire*

"The book is masterful. MacKinnon has the craftsmanship for the challenge. He uses flashback well, creates arresting images [and] has enough travel experience to take the bizarre in stride." *The Globe and Mail*

THE
ONCE
AND
FUTURE
WORLD

NATURE AS IT WAS,
AS IT IS,
AS IT COULD BE

J.B. MACKINNON

RANDOM HOUSE CANADA

PUBLISHED BY RANDOM HOUSE CANADA

www.randomhouse.ca

LIBRARY AND ARCHIVES CANADA CATALOGUING IN PUBLICATION

MacKinnon, J. B. (James Bernard), 1970–
The once and future world : nature as it was, as it is, as it could be / J. B. MacKinnon.

Includes bibliographical references.
Issued also in electronic format.

ISBN 978-0-307-36218-6

1. Human ecology. 2. Nature—Effect of human beings on. 3. Restoration ecology.
I. Title.

GF75.M23 2013   304.2   C2012-905605-7

Text and cover design by Terri Nimmo

Cover image: Paul Fuller / Millennium Images, UK

Printed and bound in the United States of America

2 4 6 8 9 7 5 3 1

For my mother
and in memory of my father;
this was the last book he ever read.

CONTENTS

Fig . 1 .

THE NATURE

of

THE PROBLEM

Somewhere in Eden, after all this time,
does there still stand, abandoned, like
a ruined city, gates sealed with grisly nails,
the luckless garden?

INA ROUSSEAU

Chapter 1.

ILLUSIONS OF NATURE

———— · ♦ ♦ ♦ · ————

P icture the first place you thought of as nature. Maybe it was nothing more than a vacant lot in the middle of a city, or a patch of scrub along a riverbank. It might have been a cottage or campground that you visited year after year, or perhaps your childhood home opened onto a forest, a beach, a mountain. Whatever your original vision of nature was, fix it in your mind.

Myself, I grew up on a prairie that had no name. I've looked into the question, hoping to turn up some lost but interesting name like those I've known from other places—Joe's Snake Field, or Our Lady of the O, or Fountain of Bones—and have come up empty-handed. The best explanation I can give for the anonymity of my home prairie is that it seemed to have hardly any history. Why give a name to a patch of grass where nothing much had happened?

Even to say it was a prairie doesn't seem quite right, because it wasn't flat or even rolling, but instead spilled down from high ridges to a river valley. Still, it was grassy and open to the sky, and in every practical sense it was infinite. My childhood landscape was the northernmost tip of the rain-shadow drylands that sprawl up most of western North America, and I could have stepped out of my house and walked three thousand kilometres to Mexico and been thirsty all the way. It was rattlesnake country and black widow country, and as a boy I was brown-skinned and blond-haired and so much a son of that sun-baked earth that I wouldn't flinch if a two-inch-long grasshopper thudded down on the bare skin of my ribs as I ran through the fields. I knew the prairie in the hands-in-every-crevice detail that only a child can, and it was, for me, a place of magic. The miracle of a mouse skeleton compacted in a pellet of owl scat! The mystery of snow flies hatching onto ice! One winter my father stopped his truck to chase down a giant, bone-dry tumbleweed that was pinwheeling in the wind. He set up that huge ball of prickles on the patio, threaded it with lights and sprayed it nightly with water until it glittered with golden icicles. It remains the most beautiful Christmas tree I've ever seen.

The fiercest animal on the prairie, and therefore my boyhood symbol of wild nature, was the red fox. The sporty, lolling, yipping red fox. It's an extraordinary animal. An adult red fox is able to run at seventy kilometres per hour. They've been observed trying to race airplanes down runways, the way dogs will chase the wheels of a car. When hunting, a fox can leap eight metres and land with enough precision to pin a mouse beneath its forepaws, meaning that at takeoff the fox has accounted for its own speed and trajectory, the speed and trajectory of the mouse, along with other factors such as wind and ground cover, *all without ever actually seeing the prey*. Such a pounce is so carefully controlled that a fox will, at times, beat its tail to one side or the other in mid-air to adjust its flight path. There were always fox dens on my home prairie.

I finished high school and, as people do, I moved away, coming home to visit ever more rarely. One day I returned to find that the nameless grasslands had finally been given a name: the Royal Heights housing development. Suburban homes now spread across the land that held my first memory of snow and of my first night in a tent alone in wild country, and of a thousand other adventures.

A small rump of prairie remained, and I went there looking for fox dens. I found none. As I walked away that day, I saw the red fox as a martyr for every harm ever done by humankind against the wild, an icon of the ceaseless retreat of fang and claw and the relentless advance of the bloodless and tame. Every year more grasslands were erased to make way for lawns or shopping centres, with the fox gradually disappearing from the unsung hills as surely as the buffalo once vanished from the Great Plains or the whales faded from the sea. My childhood home had become my lost Eden.

Just about everyone on earth, I suspect, has his own version of this same story—the childhood wilderness despoiled. For me, it was the beginning of a journey that would change the way I see the natural world. I came to realize that we, you and I, cannot hope to make sense of this thing we call nature by looking at what surrounds us, or even by seeking the wilderness. Instead, as science has begun to recognize, we need to reach back and revisit the past—tens, hundreds, even thousands of years ago. What we find there is the living planet at its most extraordinary, often so far beyond what we know today that it challenges our expectations of what life on earth can be. The good news is that time travel is just the way we imagine it, full of marvels and surprises, odd beasts, ancient mysteries, and lands that have never known a human footfall. But the history of nature also takes courage. It calls on us to remember losses, not only in the wild, but within ourselves. The past asks us how, what and why we allow ourselves to forget.

When I began to look into the story of the foxes, I expected to uncover the usual sad chronicle of decline, another species vanishing point by point like stars obscured by city lights. Instead, I learned that the foxes of my youth had trotted onto the scene only a few decades ahead of my own arrival as a five-year-old boy—that they were, really, not much more a part of the natural order than the housing development that had displaced them. In fact, if you live in North America and have ever seen a red fox, have ever taken some delight in the briskness of its movement and intelligence of its expression, then what you have seen is almost certainly an animal that is not a part of the native wildlife.

When the first Europeans to settle in North America arrived on the east coast, they found themselves in a land apparently devoid of the red fox. Beginning in the 1700s, they began to import the animals so that they could pursue them for sport as they had done back home, in English-style horseback hunts. Some foxes escaped and, like the European colonists themselves, began to drift westward. People later introduced red foxes in other corners of the continent, accelerating their spread.* By

---

* There's an apocryphal story from my hometown of a fox-fur farmer who thought he could save on the cost of fences and cages by setting up on an island in a huge inland lake. Then a winter cold snap froze the lake's surface, setting the foxes free.

about the 1980s, the canine known to science as *Vulpes vulpes* had taken over North America from east to west.

Biologists consider the red fox an invasive species—they can do serious harm when they move into a natural system they were not a part of before. Red foxes threaten some two dozen rare animals in California, including such federally endangered species as the Santa Cruz long-toed salamander, least Bell's vireo, blunt-nosed leopard lizard and giant kangaroo rat. Introduced red foxes have caused major declines in many of Australia's wonderfully named small beasts, from rock-wallabies and brush-tailed bettongs to quokkas and numbats. They can spread diseases such as rabies, distemper and mange. Not every introduced species is a problem, but the red fox makes the top 100 list of the world's worst, as compiled by the Global Invasive Species Database.

In places, introduced red foxes have even driven native fox species off the landscape, and this is where the issue becomes confusing. As it turns out, North America actually *was* home to red foxes before Europeans introduced the animals, but these native foxes were adapted only to northern boreal forests and to certain mountain ranges in the west. I hoped to discover that I grew up among native foxes, but biologists considered it unlikely and every clue I turned up suggested that foxes were not present in the past. In the 1860s, for example, a pair of British immigrants began hosting English-style hunts over the sagebrush hills around my hometown. The events were complete in detail right down to the imported hounds and the cries of tally-ho. Only one part of the fox hunt broke from tradition: they didn't hunt foxes. Finding none in the area, the hunters pursued coyotes.

To learn that the red fox, my personal symbol of the wild, was not deep-blooded on the landscape, not "natural," felt like a blow to my sense of self. I told one of my brothers what I had discovered and he said, "I don't believe you." I began to reel off the evidence and he said, "Nothing you can say will make me believe you." You think you know the truth of a place, and then—you don't. The fox, however, was only the beginning.

♦ ♦ ♦

Nature is a confounding thing. The question of whether humans are a part of nature or stand apart from it has probably been debated since the days of the first campfire. In one sense, the answer is obvious enough: whether you believe human beings are the result of a lengthy process of evolution or a sudden act of divine grace, there is no question that we are flesh-and-blood animals, carbon-based life forms spun from the same celestial dust as the rest of creation. At the same time, we have always sought to define ourselves as separate from all other species, whether through our capacity for self-awareness and rational thought or the presumed existence of the human soul. Such efforts often carry more than a whiff of desperation: one philosopher saw a sign of human exceptionalism even in the fact that our noses are a "marked projection" from our faces—apparently unaware of, say, the proboscis monkey, which has a hugely bulbous nose that dangles down to below its mouth.

The same contradictions pervade our relationship to the landscape as a whole. When the German biologist Ernst Haeckel sat down in 1866 to give a name to the study of nature's systems, he began with the ancient Greek word for "house," *oikos*, and coined the term "ecology." The living planet is our home. As of 2008, however, a global majority of people live in cities, where that idea is increasingly distant and abstract. We're surrounded by a world that is, by our own description, "man-made" and "artificial"; nature is what rises up at the edges of cities and towns, or wherever else it has not been beaten back by human hands. We often put the two—natural versus unnatural—in opposition, weighing whether or not to preserve the former or make way for the latter, all the while assuming we can distinguish one from the other. This is nature by our most ordinary definition: the sum total of everything that is not us and did not spring from our imaginations.

We're all aware that dramatic events have played out in the human relationship to the natural world, from the extinction of the dodo to the collapse of the Atlantic cod fishery to the logging of the Amazon rainforest. Yet nature also seems somehow outside of history, always expressing itself in a weed poking up from a crack in the concrete—give it half a chance and it will erase our puny imprint as surely as it buried the pharaohs of Egypt in desert sand or the Mayan temples in jungle vines. We stand on a stretch of wild seashore or see a mountain covered with trees and make what the cognitive psychologist Gary Marcus calls "the error of the historical present"—we assume that it is as it always was, at least by any measure of time that we can grasp. Nature was here long before we were, and will linger long after we're gone.

As a field of study, the history of nature is remarkably young. It doesn't even have a settled name. "Natural history" was taken long ago as a catch-all for the natural sciences (botany, geology, paleontology, etc.), leaving us to struggle with descriptors like "environmental history," "ecological history," "historical ecology" and "green history." None of these terms is even fifty years old, and the bulk of the research is more recent than that. The first ecological history of North America—*The Eternal Frontier*, by Tim Flannery—was printed in the twenty-first century.

To find the roots of the science, it's useful to look back to 1864, when the New England scholar George Perkins Marsh published a book titled *Man and Nature*. Marsh had an interesting life: self-described as "forest born" on a Vermont homestead where wolves and mountain lions still prowled, he went on to read and write in twenty languages, decide the final dimensions of the Washington Monument, and act as America's ambassador to the Ottoman Empire. Marsh lived at a time when entire new civilizations were being founded in places like Australasia and the Americas, while the riches of these regions renewed the empires of Asia and Europe. When Marsh wrote *Man and Nature*, popular belief still held that the natural wealth of the earth was infinite. There was not much reason to think otherwise: in North America at that time, for example, buffalo in the hundreds of thousands still roamed the Canadian Prairies and Great Plains, grizzly bears skulked through every mountain range in the west and more than half a century would pass before the last million-pound hauls of shad would be fished from the rivers of the Eastern Seaboard. Marsh's accomplishment was to see what others of his day could not.

Marsh was the first to popularize the idea that humankind is not the righteous redeemer of nature's bounty, but instead a disturber of natural harmonies and a threat to life on earth. To today's reader, that message—along with the book's apocalyptic tone—is instantly familiar, if not hackneyed. At the time, it was shocking. *Man and Nature* made a major contribution to the rising interest in wilderness preservation that took hold in the late nineteenth century, leading to the establishment in 1872 of the world's first national park, Yellowstone, which today includes parts of Wyoming, Montana and Idaho; the book directly inspired the creation of Adirondack Park, which remains the largest protected area in the contiguous United States. Marsh's relatively scholarly writing, however, was soon overshadowed by more poetic contemporaries such as Henry David Thoreau and John Muir, whose celebration of American wilderness has set the tone of conservationism ever since as a battle to defend pristine nature from human degradation.

What was forgotten—what the world was perhaps not ready for in Marsh's time—was *Man and Nature*'s more subtle point. Travelling the world 150 years ago, Marsh concluded that most of the planet was not a threatened wilderness, but had already been "much modified in form and product." He notes, for example, that the papyrus plant used to make writing paper by the ancient Egyptians had, by his own era, nearly vanished from the Nile River. In Ravenna, Italy, he measured the doors of an ancient cathedral and found they were made of grapewood thicker than any vine that still existed.* He recorded that ostriches had once lived not only in central and southern Africa, but all the way north to the Mediterranean and across the Arabian Peninsula three thousand kilometres to the nation of Oman. Studying the histories of ancient armies, Marsh noted that many of the most arid and abandoned landscapes of Europe and Asia were once so fertile that large armies made long marches through them on local food supplies alone. In his own lifetime Marsh had seen earthworms, introduced to New England from Europe, go from being so rare that anglers kept secret the few locations they could be found, to so numerous that some freshwater springs

---

* The planks measured four metres long by half a metre wide, which Marsh said was far larger than any living vine he could find, "though I have taken some pains on the subject."

were soured by the taste of their rotting bodies. He made note that many of the oak forests first witnessed by pioneers on the east coast of North America had been maintained in a "park-like" fashion by indigenous nations, and that seals used to visit the fresh water of Lake Champlain on the border of Quebec and Vermont, where they are utterly unheard of today. Marsh was the first to promote the now widely held idea that the collapse of the Roman Empire was in large part an ecological collapse—"rivers famous in history and song have shrunk to humble brooklets"—and even speculated that the reason that writers from the age of classical antiquity failed to remark on the phenomenon of ocean phosphorescence was because they never witnessed it: that civilization since then had transformed the seas to such an extent that luminescent plankton were freed to set the waves aglow.

Man and Nature is today considered one of the founding texts of the environmental movement, and there's no question that Marsh argued for the preservation of the wild. But he also went further. More than a century before the first Earth Day, Marsh was calling for the renovation, literally the "making new," of an exhausted planet. His concern was not first and foremost how much land to set aside here or there in parks and protected areas—the question that launched a hundred years of battles for the world's last and best wild spaces. Instead, Marsh asked whether we were changing nature itself into something new, something lesser, something our ancestors might not even recognize. He had written what can be thought of as the first principle of historical ecology: to know what is, you must know what was.

♦ ♦ ♦

That isn't as easy as it might sound. I mentioned earlier that my childhood landscape seemed to have hardly any past, and this is true in a textbook sense. The first European to visit the area didn't arrive until 1811. By that time, there was already steam-powered ferry service between New York and New Jersey, London had a population of more than one million, and Captain James Cook had circumnavigated Antarctica. The place where I grew up never saw war or revolution, was never dark and bloody ground. George Washington's false teeth have a longer written history than my hometown.

But then, no ordinary telling of history commemorates the caribou. Most people know caribou as the strange, bone-jutting beasts that survive on Arctic barrens, yet caribou were hunted within living memory on a plateau across the valley from my family's kitchen window; it was probably there that a prospector in 1926 witnessed the animals being "wantonly slaughtered." My parents worked at Cariboo College,* but not once while growing up did I hear that the name referred to the former presence of the actual animal.

No one ever spoke of the elk, either, described in one account as "numerous and widespread"—their antlers were still bleaching in the hills as the twentieth century began. An early naturalist wrote that the Western rattlesnake could be found on some hillsides "coiled upon every ledge, stone and bare spot"; I saw just two rattlers in my entire childhood. The schools I went to never taught that the sage grouse, white-tailed jackrabbit, pigmy

---

* I remember the devastatingly of-the-era CARIBOOZER sweatshirt, which featured the image of a drunken caribou; the college has since been renamed, erasing even this tentative connection to the past ecology.

short-horned lizard and viceroy butterfly, among other species, had vanished from the hills. I knew that bounty hunts had helped to eradicate the wolf from much of North America, but had never been told that the landscape I lived on shared that history, or that the killing had gone far beyond wolves to include any animal that might steal a chicken, eat a kernel of wheat or otherwise intrude on human interests, from owls and eagles to sparrows and prairie dogs. Some were massacred in "ring hunts," with the spirit if not the huge breadth of one recorded from eighteenth-century Pennsylvania, when a man named Black Jack Schwartz organized two hundred settlers to circle an area fifty kilometres in diameter and then close in on every bear, buffalo, elk, deer, cougar, wolf, bobcat, wolverine, fisher, otter and beaver in the area—allegedly nearly a thousand animals in all.

"The gun and the plough, the saw and the cow, the dam and the ditch"—so goes one abbreviated history of my hometown. By 1850, settlers' cattle and cayuse horses had grazed whole horizons of grassland to the ground. Feed was shipped in, and with it came the seeds of the invasive plants—Kentucky blue-grass, smooth brome, diffuse knapweed, crested wheatgrass—that swept through the native prairie to become the familiar vegetation of my youth. The Christmas tumbleweed I remember so fondly was a stowaway from Russia.

Earlier still was the fur trade, which drove an annihilation of wild mammals so total that trapped-out regions were sometimes called "fur deserts." A typical pack train of the era involved three hundred horses, each loaded with twin eighty-pound bales of pelts, for a total of forty-eight thousand pounds of furs. The Scottish naturalist David Douglas, apparently outraged by a land-scape empty of any furbearing animal larger than a chipmunk, told the chief trader of the fort that went on to become my hometown that "there is not an officer in it with a soul above a beaver skin."

Even the fur trade is not the beginning of the depletion. Millennia of indigenous settlement came before that, the tribes and cultures that, by the time they encountered men of European descent, were so thickly populated on the landscape that one early visitor reported he was never out of sight of campfire smoke. In one local legend, Buffalo charges at Coyote for disrespectfully kicking at the bones of Buffalo's ancestors. "They were all killed, and my brothers and sisters were taken away," Buffalo says. "I am the only one left." The archaeological record shows that buffalo, also known as bison, roamed at least near enough to my home prairie that a herd could have covered the distance in a weekend. Never once while growing up did I hear that the land might have been buffalo country.

In the year 2000, a local biologist named David Spalding, attempting to express in simple terms the reason so many species had disappeared, put it down to "the general increase in humans." The soil itself was once crusted with a living skin of lichens and micro-organisms, hundreds of years in the making and now trampled into powder by livestock in most places. One such cryptogamic crust species, *Diploschistes muscorum*, is a living irony: it's easily destroyed by grazing cattle, but looks like a cow-pie. The resemblance generates Abbott and Costello dialogue such as this, from a grasslands walk I took with a local conservationist: "There's some lichens! Or maybe that's just cowshit." Today, a dubious milestone has been reached. In this overlooked corner of the globe, this place that seems so empty of history, the natural world has undergone so much change in just the past one hundred years that a person like me, raised so close to the land that my feet were stained the colour of it, wouldn't feel at home in the original grassland. The prairie as it was is an unfamiliar country.

Yet if I took you onto the remaining grasslands around my hometown today, they would seem to you as ancient and unchanging as anywhere on earth. You'd smell sage and the vanilla scent of ponderosa pine bark, and you'd hear meadowlarks, and the bunchgrass would rustle like the restless dead. Cicadas would zing in swales of aspen, and if you were lucky with the year and the season, the brittle prickly pear would be in lemon-yellow bloom. The breeze would suck across the hilltop balds where it has carried away the soil and leave you blinking as it dried your windward eye. In the dust you might see stripes where gopher snakes had lain to warm themselves in the morning sun. You might see an ant lion spitting sand as it tries to knock insects into its pit trap. You might see fox tracks. The whole of the landscape, from sky to soil, would have the look and smell and feel of what we call nature. It is an illusion that has in many ways created our world.

Chapter 2.

KNOWLEDGE EXTINCTION

——— · ◆ ◆ ◆ · ———

In April 2010, the *Deepwater Horizon* oil rig sank in the Gulf of Mexico, leaving behind a gushing oil pipe one-and-a-half kilometres under the sea. The spill is already fading from memory. At the time, though, as the weeks passed and the disaster continued, the world watched each day's new round of coverage with horror: oil-soaked pelicans, poisoned mangrove forests, debates about whether or not to set the ocean's surface on fire. Two months into the calamity, U.S. President Barack Obama declared that the cleanup of the spill would go beyond the "crisis of the moment." He promised a long-term recovery plan that would return the Gulf Coast and its waters to "normal."

When it comes to nature, normal is in the eye of the beholder. The discovery of this fact—perhaps the most important in our relationship to the living world—has its roots in an unexpected discipline: child psychology. In 1995, Peter H. Kahn Jr. and Batya

Friedman, both of Colby College in Maine, published a study of the environmental views and values of children from a mainly poor, mainly Black community in Houston, Texas. Houston, located on the Gulf of Mexico's coastal plain, is one of America's more polluted cities, yet only a third of the kids reported that environmental problems affected them directly. In an attempt to explain this unexpected outcome, the authors write:

> One possible answer is that to understand the idea of pollution one needs to compare existing polluted states to those that are less polluted. In other words, if one's only experience is with a certain amount of pollution, then that amount becomes not pollution, but the norm against which more polluted states are measured . . . Indeed, what we perceive in the children we interviewed might well be the same sort of psychological phenomenon that affects us all from generation to generation. People may take the natural environment they encounter during childhood as the norm against which to measure pollution later in their life. The crux here is that with each generation, the amount of environmental degradation increases, but each generation takes that amount as the norm.

Ideas often arise simultaneously in disparate quarters. The same month that the child development study was released, fisheries scientist Daniel Pauly published a commentary about what he called "shifting baseline syndrome." Pauly had been inspired, in part, by the 1984 book *Sea of Slaughter*, in which author Farley Mowat reviews five centuries of explorers' journals and pioneer accounts to expose the terrible toll of human hunting and fishing in the North Atlantic. The book had recently been revisited by three biologists who concluded, based on Mowat's research, that biomass—the total weight of living things—off North America's east coast may have declined by 97 percent since written records began. The failure of coastal residents and scientists to recognize such a shocking diminution seemed to Pauly explainable only by a long-term pattern of amnesia. Each generation of people saw the coast that they grew up on as the normal state of nature, and measured the declines of sea life against that baseline. With every new generation, the baseline shifted—"a gradual accommodation of the creeping disappearance," Pauly said. We were forgetting what the world used to look like.

Researchers have since found evidence of shifting baseline syndrome among people and in places as wide ranging as bushmeat hunters in Equatorial Guinea, birdwatchers in Yorkshire, England, and villagers along the Yangtze River in China. As chance would have it, one of the most startling studies was carried out in the Gulf of Mexico and adjoining Caribbean Sea. Marine biologist Loren McClenachan, currently an associate professor of environmental studies at Colby College, compared big-game fishermen's photos from the Florida Keys from the 1950s to the modern day. In the old black-and-whites, the biggest fish, strung up on the dock, are as tall and wide as the fishermen themselves, while the rest of the day's catch—the fish have an average length of nearly one metre—is piled up in heaps. By 2007, the catch is dominated by snappers that measure just a little longer than a grade-school ruler; the "small" fish of past years are often larger than the trophy fish of recent times. Most striking of all is that the fishermen look equally pleased with themselves through the generations—the same wide smiles, the same backslapping-with-Hemingway pride.

Many of today's fishers responded to McClenachan's research with flat disbelief.

Another of McClenachan's studies found that coral reefs across the Caribbean and Gulf of Mexico once were home to at least four tonnes more fish per hectare than they typically are today; to put that into perspective, four tonnes of fish is enough to feed a meal to twelve thousand people. To see those coral reefs in their glory, we'd need to reach back to the seventeenth century, when the waters were also home to an estimated 300,000 Caribbean monk seals (now extinct, the last confirmed sighting in 1952); or centuries earlier still, when as many as 91 million green sea turtles churned the waves (today's numbers are less than 1 percent of that figure).

Perhaps the most remarkable research involves two humble varieties of sea sponge, once so significant a part of the aquatic environment that in the first years of the twentieth century some twenty thousand tonnes of sponges were hauled ashore every year in Florida and the northern Caribbean alone. In 1939, the wild sponge population, already seriously depleted for uses ranging from household scrubbing to birth control, was hit hard by epidemic disease. It has never recovered. Sponges have an astounding capacity to remove microbes from water; in a single day, a sponge the size of a soccer ball can sieve 90 percent of the bacteria from more water than you will drink in your lifetime. The loss of the sponges damaged water quality throughout the Caribbean and Gulf of Mexico, resulting in a crash in the number of lobsters, and of an economy that had sustained thousands of sponge and lobster fishers.

When McClenachan goes diving in those waters today, she no longer sees with the same eyes that she used to. "It's like there's all these ghosts lurking around," she says. She has become, she tells me, "a big buzzkill for family vacations."

♦ ♦ ♦

Memory conspires against nature. The forgetting can begin in the instant that a change takes place: the human mind did not evolve to see its surroundings—what we now so clinically refer to as "the environment"—as the focus of our attention, but rather as the backdrop against which more interesting things take place. We generally don't notice small or gradual changes. Our minds would otherwise be crowded with turning leaves and the paths of clouds across the sky—a beguiling madness, but a madness all the same.

Even a dramatic event can be overlooked in the moment, through a phenomenon known as "change blindness." In the most famous study of its kind, test subjects who were asked to follow the path of a ball being passed among a group of basketball players consistently failed to notice when a person in a gorilla suit danced through the scene. It's a question of where you direct your attention: keep your eyes on the ball and you're likely to miss the dancing gorilla. While being guided through a similar demonstration by Daniel Simons, a University of Illinois psychologist, I failed to notice a slow but steady change to the background of a scene—despite the fact I was aware this was the purpose of the video. In fact, I missed the change three times straight. Then Simons patiently suggested that I run the video in fast-forward, which made the change abrupt enough that I finally saw that about one-fifth of the wheat field I was looking at was being slowly reduced to stubble.

Most people do not believe that such experiments would fool them. In reality, change blindness affects even the expert eye. A study of football fans found they were 110 percent more likely than non-fans to spot changes to a football scene—but only if those changes were relevant to the game. Make a change in the background, and both fans and non-fans were likely to miss it. Yet the belief that your own eyes will not fool you is persistent enough that psychologists have given that condition a name too: "change blindness blindness." If you don't believe that you are capable of missing significant changes to a scene, then you won't heighten your awareness in order not to miss them— which means that you probably will. Change blindness blindness is the failure to see that we so often fail to see.

We are an incredibly adaptable species. Whether or not we notice a change in our circumstances, the change itself is real, and we quickly adapt to the new conditions. Once we've done so, there is little point in holding on to memories of how things used to be. The shifting baseline syndrome applies as much to the way we forget what houses cost ten years ago or fail to notice that fast-food portion sizes have tripled since the 1970s as it does to the natural world. Out of sight, out of mind: ordinary amnesia.

The historian Clive Ponting sees the whole story of civilization as a series of adaptations forced on us as we rendered the world around us less and less livable. What he calls "the first great transition"—the era in which our ancestors moved over several millennia from being hunter-gatherers to being agriculturalists—is an example:

Human societies did not set out to invent agriculture and produce permanent settlements. Instead a series of marginal changes were made gradually in existing ways of obtaining food as a result of particular local circumstances. The cumulative effect of the various alterations was important because they acted like a ratchet. Changes in subsistence methods often allowed a larger population to be supported but this made it difficult and eventually impossible to return to a gathering and hunting way of life because the extra people could not then be fed.

Ponting argues that this "ratchet effect" has continued, with societies needing to advance their technologies and degree of organization in order to respond to environmental challenges that are often of their own making. Success has the paradoxical effect that even greater human populations with larger environmental impacts can support themselves from a more degraded natural world. The author Ronald Wright describes this as a "progress trap" that contributed to the collapse of societies as significant as the Roman and Mayan empires.

The adapt-and-forget pattern is amplified by modern life. If you, like me, are a city dweller, then you're unlikely to suffer change blindness to shifts in the natural world, because you're not there to witness those shifts, and you don't suffer much environmental amnesia, either, because you don't have many memories of nature in the first place. For you, the baselines that shift will be mainly urban and technological ones; your generation will accept as normal that which your parents struggle to adapt to, and your children will carry forward little memory of the city as you knew it.

Memory is depressing country, and never more so than when it comes to what is passed down from one generation to the next. Much of the research into such memories has to do with the Holocaust. There are urgent reasons to remember the Nazi genocide against Jews and other selected minorities in Europe during World War II, most obviously in order to prevent anything similar from happening again. Yet Holocaust researchers have had to confront difficult truths about remembrance, including the fact that survivors' stories will be largely forgotten by the time they've been handed down through just three generations, or about ninety years—the great-grandchildren of the men and women who have personal experience of the Holocaust will, like the rest of us, know the event mainly from books and films. The Holocaust is drifting toward "remote history," or history that was never directly experienced by any living person who can remember it. Much of what happened, how, and why, will be known only to a tiny cadre of specialists, while many more details—each a part of the historical caution that the Holocaust represents—will be irretrievably lost. The term for this is knowledge extinction.

Oddly enough, the extinction or near-extinction of certain animals has proved to be the standout exception to our forgetfulness when it comes to the natural world. The extinct dodo, for example, has become an unlikely cultural heavyweight. A fat, flightless bird that disappeared ten generations ago, the dodo ranks alongside the penguin, elephant and tiger among animals that even small children are likely to recognize. Popular culture has also held on to the bison—our greatest and most enduring symbol of the natural abundance of the past. The image of that great shaggy head immediately calls to mind the thundering herds that ruled North America's plains two hundred years ago, and also the destruction of those herds, the buffalo hunters firing until their gun barrels seized from the heat of it, the skulls piled high as houses.

The dodo and the bison have passed into the ranks of what are known as transgenerational memories: stripped-down versions of the original that can be recalled whenever cultural shorthand is needed to represent some era or moral or way of being in the world—Marilyn Monroe and the Berlin Wall and the Temptation in the Wilderness. A transgenerational memory is better known as a myth, a fable, a testament, an icon.

We hang on to the dodo, then, but lose sight of the long list of other species that disappeared from Mauritius, the Indian Ocean island the dodo called home, among them the Mauritius scops-owl, the Mauritius giant skink, the Mauritius blue-pigeon, the lesser Mascarene flying-fox, two kinds of giant tortoise and a parrot that might have been the largest ever known. We remember the bison herds too, but not the way the spadefoot toad and western chorus frog sang from the animals' wallows—an estimated 100 million small ponds across the plains—in spring, or that birds such as the McCown's longspur and the mountain plover nested on those same wallows as the sun dried them into dust bowls. The bison of memory is forever an animal of the grasslands, and never of the Rocky Mountain passes or Mexican deserts where they also once lived. Who remembers the menacing, coal-black bulls hunted in the hardwood forests of Pennsylvania? What about the bison herds of California?

Certain kinds of memorial encourage forgetting. At the time of the great buffalo hunts, almost every other animal in the New World was also under assault, from oysters to fur seals, from prairie-chickens to basking sharks. We remembered the destruction of the bison as a way of remembering that entire era of extermination; today, only the buffalo's story is universally known. The bison hunt has passed through the irony machine of history, in much the same way that the enduring presence of the Holocaust has ended up overshadowing the remembrance of every act of ethnic cleansing before or after.

To more fully appreciate the lost memory that our focus on the buffalo represents, consider the fact that a similar slaughter had already taken place in North America by the time the bison hunts began: deer were once hunted to the brink of extinction. Try raising this around a kitchen table in the U.S. or Canada, and you will meet with flat disbelief. Deer? Deer eat tulips in suburban gardens. Deer show up in online videos goring people's dogs within city limits. The drivers of New York state alone run down seventy-five thousand deer a year.

The slaughter of the deer is vanishingly obscure. Fur-trade scholar Charles Hanson declares the buckskin trade of the American southeast "sadly neglected in literature"—and the buckskin trade of the American southeast is by far the best known of the deer hunts across North America. Early colonial immigrants to the Carolinas and Florida reported plains and forests "crowded with deer." An observer named Thomas Ashe, in the 1680s, tells of "deer of which there is such infinite herds, that the whole country seems but one continuous park." To the Muscogee Indians, whose territories covered much of what is now Georgia, Alabama and northern Florida, the herds were the currency of survival. Three quarters of the Muscogee meat supply came from white-tailed deer, not to mention most of their clothing, housewares, tools and such distinctive paraphernalia as flutes hollowed from the leg bones.

Not surprisingly, skilled Muscogee hunters quickly became the supply side of the deerskin trade. On the demand side was all of Europe, where deer had already been so badly overhunted that gloves in Paris were reportedly being made with rat skins. Before the era of denim, there were deer-leather breeches, and just as with blue jeans, these buckskins were worn first by labourers and then came into fashion among the aristocracy. Imagine the scale of killing today if even a single city the size of Los Angeles, London or Toronto were to replace its jeans with animal hide. When the southeastern deerskin trade peaked in the years ahead of the American Revolution, the total number of hides brought in by Indian traders was at least one million each year. So went the pattern as the hide trade spread across North America, eventually overlapping with the bison kill. By 1886, a pioneer in rural New York recalled for the *New York Times* the hunts that routinely killed forty or fifty deer at a time in that state. Those days were done. "If a man were offered a million dollars for a deer killed in this county today he could never earn the money," the woodsman said.

"They were so scarce," writes Leonard Lee Rue in *The Deer of North America*, "that their same numbers today would make them candidates for the endangered species list."

It's easy to assume that the deer hunt is forgotten because the story ends so differently than the elegies to the bison. Unlike the buffalo, deer made a phenomenal comeback. With hunting restrictions in place, their predators largely wiped out, and forests opened by roads, farms and logging, the deer population recovered; the white-tailed deer in particular is the only large animal in North America that now ranges over more territory than ever before. But the deer trade left its own deep scars. At the beginning of the buckskin era in 1685, a Muscogee hunter would undergo ritual purification before the long winter hunt, and might make four hundred kills in a season, feeding entire communities. By the end of the American Revolution, not quite a century later, the natural economy of the Muscogee was in collapse and they were no longer able to find enough "bucks"—the origin of the slang term for money—to pay off their debts to colonial traders offering easy credit, especially for tafia rum. In 1802, federal Indian agent Benjamin Hawkins spelled out the new reality in his advice to a Muscogee leader. "Sell some of your waste lands," he said, referring to territories that were seen a century earlier as infinitely rich. "I see no other resource that is very abundant." Most Muscogee were ultimately displaced west to Oklahoma; the only land the tribe still holds in its historical territory is a 230-acre reserve in Alabama. A world emptied of deer marked the beginning of a long period of dispossession for the Muscogee, a lifetime before the scorched-earth buffalo hunts became an official instrument of war against the indigenous nations of the Great Plains.

But we forget.

♦ ♦ ♦

When the bison were all but gone, buffalo hunters still hung on in the buffalo-hunting towns, waiting for the herds to return. The animals had migrated, the men told themselves, and would come back soon. Many of the hunters saw themselves as blameless, left unemployed by the whims of natural forces. They waited a while for the buffalo, and then they became cowboys.*

Denial is the last line of defence against memory. It helps us to forget what we'd rather not remember, and then to forget that we've forgotten it, and then to resist the temptation to remember. "The ability to deny is an amazing human phenomenon, largely

---

* Similarly, the birds today known as "cowbirds" were formerly known as "buffalo birds."

unexplained and often inexplicable," writes the sociologist Stanley Cohen, author of *States of Denial*. Yet we find denial useful. It fulfills, to quote the definition preferred by Cohen, "our need to be innocent of a troubling recognition."

Once upon a time, there were dodo deniers. For more than a century after the last dodo died in the late seventeenth century, the bird's former existence was doubted and rejected. The general public forgot the bird entirely, and even naturalists dismissed reports and paintings of the bird as fanciful works of imagination. It was only the publication, nearly two centuries later, of Lewis Carroll's *Alice's Adventures in Wonderland* that brought the dodo back to life, albeit as an extinct species. The illustration of the dodo in the book, drawn by John Tenniel, captured people's hearts. Even today no one can say exactly what a living dodo looked like, but the basic image has not changed since Tenniel's depiction: the madhouse eyes, the ungainly beak, the wallflower plumage, the pointless wings. The bird is a portrait of the perfect victim.

Until the early 1800s, many leading thinkers denied the idea of extinction entirely; it was considered contradictory to the notion of godly creation. "That no real species of living creatures is so utterly extinct, as to be lost entirely out of the world, since it was first created, is the opinion of many naturalists; and 'tis grounded on so good a principle of Providence taking care in general of all its animal productions, that it deserves our assent," wrote Thomas Molyneux, a seventeenth-century scholar who argued that fossilized skeletons of Irish elk—an enormous deer that once roamed much of Eurasia—were in fact only the remains of a misplaced breed of American moose. Similarly, Thomas Jefferson, in his role as third president of the United States, hoped that the overland expedition of Meriwether Lewis and William Clark, which ran from 1804 to 1806, would find living mammoths in the American West as proof that the Christian god would not allow any of his flock to disappear from the earth.

Name a vanished animal, and in its story you are bound to find denial. Sometimes, the species is said never to have lived. More often, it's said never to have died. The great auk was among the first animals to be driven extinct after the European discovery of the Americas, alongside even more deeply forgotten species, such as the sea mink. Like the dodo, the auk was a flightless bird, though one that resembled a small penguin and spent most of its time at sea. It took a thousand years for European hunters to eradicate the auk from their home continent and, with better technology, three hundred years to erase it from North America. Auks were hunted for their meat, eggs and feathers, as well as for "trane-oil"—the animal oil that lit and lubricated the world ahead of the petroleum age. Auks were so fatty that there are reports of them being thrown, sometimes alive, onto fires as fuel to boil the oil out of other auks.

As the great auks vanished, the nature of the birds changed in the eyes of human observers. In the beginning they were seen as too thick-headed to flee from hunters, and so common that an auk was a ship captain's first sign that he was nearing the North American coast. Once the birds had become rare, their absence was blamed on natural timidity or, in one example, an alleged migration to the Arctic "by choice and instinct." At last, only a few years after the last auks were killed in 1844,* one

---

* The last known auks were a breeding pair killed on Eldey Island, Iceland; it's said that their egg was broken in the melee.

commentator wrote that "in all probability, the so-called great auk of history was a mythical creature invented by unlettered sailors and fisherfolk." Even in the 1960s, when the prior existence of the great auk was no longer disputed, a Canadian fisheries bureaucrat told news reporters that the bird had been a relict species with no place in the modern world. They "had to go," the scientist said.

But to witness the lengths to which the hand-washing rituals of denial can go, consider *Thylacinus cynocephalus*. To begin with, it was denied the uniqueness of its being. The animal is sometimes remembered as the Tasmanian tiger, because it had stripes, and sometimes as the Tasmanian wolf, because it had the pointed snout and long-trotting look of a canine (charming detail: its ears remained erect even when the animal was asleep), and sometimes even as the Tasmanian hyena, because it didn't really look like either a cat or a dog. What it was, was itself: a marsupial that carried its young in a pouch, like a kangaroo,* but was otherwise a formidable hunter and meat-eater with no close relatives among the living creatures of the earth. Most biologists now refer to the animal as the thylacine.

By any name, the species had its true nature cast aside at every turn in order to aid and abet its destruction. Thylacines originally lived both in Australia and on the island of Tasmania. By the time European sailors visited these places, the thylacine had already faded from the continental mainland, possibly as much as three thousand years earlier, and probably due to changes in the ecology caused by the arrival and enduring presence of human beings—the people who came to be known as

---

* Male thylacines had pouches also, which they used to stow their scrotal sacks.

Aborigines. On Tasmania, however, Aborigines and thylacines managed to live alongside one another, and the animals were abundant enough that even the first party of Europeans to set foot on the island encountered thylacine tracks. Shy, cryptic and most active at night, the "Tasmanian tiger" was, at first, mainly a mystery to the colonists. As European-style fields and farms spread over the Tasmanian landscape, however, reports began to spread of thylacines killing sheep and chickens. The losses appear never to have been very great, but the thylacine was quickly made into a monster. Where once a settler might write of feeling "very lucky to have been so close to a tiger" or remark that, in Tasmania, "there is nothing that will hurt a man but a snake," suddenly the thylacine was so feared and hated that men who killed one often burned its skin and smashed its bones. The idea of dying in the fangs of a thylacine took on a nightmare quality—the animals were said to kill like vampires, draining their victims of blood. Having gained supernatural powers through human storytelling, the thylacine was denied its flesh and blood vulnerability. By the late 1800s, when scientists were having difficulty finding any thylacines at all, sheep ranchers still claimed the hills were "infested" with them.

Science, meanwhile, promoted its own form of denial. Struggling to explain why many of Australia's distinctive species, including the thylacine in Tasmania, appeared to be rapidly declining in the face of European settlement, many academics were quick to blame the animals themselves. They were "the stupidest animals in the world," "a race of natural born idiots," wrote the superintendant of the National Zoological Park in Washington, D.C., in 1903. Other respected scholars declared the thylacine "sealed off . . . from the great evolutionary advances that took place elsewhere on earth," "badly formed and ungainly and therefore very primitive," "unadaptable and so ill-fitted for survival in a changing world." In 1936, when the thylacine went extinct after a century of habitat loss and extermination campaigns, another brand of denial was added: the claim that no one had known the animals were so close to the brink. In an exhaustive review of the evidence for that enduring belief, Australian researcher Robert Paddle uncovered at least twenty-five warnings about the scarcity of the thylacine. In fact, the species was protected under law exactly fifty-nine days before the last known thylacine died in captivity. "The pathetic excuse—'we did not know what was going on'—was as unreal in Tasmania in 1888, as it was in Germany over fifty years later," writes Paddle. "The sense of loss associated with extinction, and the acceptance of responsibility, guilt and blame for allowing it to happen, are not easily borne."

The last thylacine is believed to have been captured in 1933 in the Florentine Valley of southwestern Tasmania. The animal was trussed and tied to a pack horse, then taken to the town of Tyenna, where it was put in a cage and briefly ogled—ladies and gentlemen, the fearsome Tasmanian tiger!—by the locals. Then it was sent to the state capital, Hobart. On arrival at that city's zoo, it was filmed for sixty-two seconds by the zoologist David Fleay, and then, in the species' most memorable final act, it bit Fleay on the ass.

Those sixty-two seconds of footage make for uncomfortable viewing. The dodo, the great auk—they're remembered only in oil paintings that bring to mind long-ago times beyond our understanding. The simple fact that you can watch a Tasmanian tiger on film is a constant reminder that the animal still existed in living memory; it had been gone for only a month when my own father was born. Behold the last thylacine. It blinks, it paces—it yawns! Its gape is astonishing, nearly the height of the animal itself and lined with slender teeth. Its eyes are anxious and curious and do not suggest stupidity. And it is extinct. The Tasmanian tiger will never blink or pace or yawn or look out at the world with anxious eyes again.

The last thylacine lived out its brief remaining days in a wire-topped cage, its only shelter a tree that was not yet in leaf so early in the antipodean spring. In the last two weeks of the animal's life, nighttime temperatures dropped below freezing, and soared as high as 42 degrees Celsius by day. One witness reported hearing the thylacine's ululating cry of distress. It died in the night on September 7, 1936. In an unsettling act of complacency, the body was apparently thrown away, though the last thylacine—only recently confirmed as a male—has since come to be remembered fondly as Benjamin, Benjy for short. From that moment onward, the species has been subjected to the only act of denial still available: it has been refused the finality of extinction. After seventy-seven years without hard evidence of a living thylacine, people still regularly claim to see the animals in the wild.

Throughout all of this dark history, one fundamental truth about the thylacine had to be constantly pushed aside. Amid all the ignorance and excuses, the blood, poison, traps and lies, a slender countercurrent has always existed in the story of humankind versus *Thylacinus cynocephalus*. From the very first, reaching back into the days of the Aborigines, there have been people who, for their own reasons, never did buy into the fear and hatred directed toward the animal. It is because of these rare few that we know that the thylacine, like other dog-like species around the world, had the capacity to live not only alongside us, but among us. With appropriate human care, they were capable, as Paddle puts it, of "perception and friendship," of "love and interest." A thylacine, in other words, made an excellent pet.

Chapter 3.

A TEN PERCENT WORLD

——————— ᐧ ◆ ◆ ◆ ᐧ ———————

S everal years ago, a grey whale swam into the heart of
Vancouver, British Columbia, a metropolis of two million
people. Such encounters are not unheard of; one year earlier, a
whale swam beneath New York's Verrazano-Narrows Bridge,
the gateway to the city's harbour, and briefly put on a show
for the residents of Staten Island and Brooklyn. Vancouver's
whale, though, passed under three bridges and straight into the
urban core up a dead-end channel narrow enough to be known
as False Creek. There, ringed by marinas and reflected in glass
towers, the whale thrilled people from every walk of life.

Most people considered that day a fluke—a once-in-a-
lifetime experience. Maybe it was. But little more than a century ago, grey and even humpback whales—the species famous
for its beautiful, mournful underwater songs—were an ordinary presence off Vancouver's shores. Hundreds of whales

lived in or passed through the area; a newspaper from 1869 reports that they "spouted their defiance" at the growing city from its waterways. They did not defy for long. Within forty years, every one of the region's great whales had been hunted and killed. Only a tiny minority of people are aware of this history, and for them, the grey whale's visit had a different meaning. It raised the hopeful possibility that what was once, may be again.

The way you see the natural world around you determines much about the kind of world you are willing to live with. If you are aware that whales once swam in your local waters, then you can ask yourself whether they might belong in those straits and bays once again. If you're unaware of the animals' past presence, then their absence will seem perfectly natural, and the question of whales in the future simply will not occur to you. Seattle, Washington, is less than two hundred kilometres down the Pacific Coast from Vancouver and shares its whaling history. There, a poll found that more than 70 percent of residents consider their local waters—which have been affected by human activities for millennia, and have suffered dramatic declines in sea mammals, fish, shellfish and seabirds—to be in good condition and not in need of restoration.

Every corner of the planet has been touched by human influence—in order to understand our current state of nature, we need to look at the present through the lens of the past. To do so is obviously complicated, involving not only science and statistics, but also the way that we experience the natural world—the sights, sounds and sensations of life on earth. Yet if this blue-green globe was once a richer and more varied place, we should be able to make some rough measure of the amount of change that has occurred. Nature as we know it today is a fraction of what it was, but what might that fraction be? No single study has made the calculation, but an accumulation of research on the decline of species after species, of living system after living system, does point toward a figure.

We live in a 10 Percent World.

Consider that just 14 percent of the earth's terrestrial surface is currently protected from human exploitation, along with approximately 1 percent of the oceans. Yet even those numbers give too optimistic an impression of what has been set aside. We have, for example, preserved nearly 40 percent of the world's snow and ice—more than any other major habitat type—but just 5 percent of its temperate grasslands, which alone cover one-tenth of the planet. Similarly, almost 30 percent of coral reefs are within marine protected areas, but only 6 percent have *effective* protection from threats that range from pollution to overfishing. Eighty percent of reefs had their largest animal species depleted even before the year 1900; rounding to the nearest whole figure, the number of coral reefs worldwide that are considered "pristine" is zero.

The planet is still home to a surprising amount of wild country. A full 44 percent of the earth's land is composed of areas larger than ten thousand square kilometres that have fewer than five people per square kilometre. But again, that figure does not represent anything like the world in all of its variety. Most of those big, empty spaces are in the polar regions, northern forests, the Amazon and the world's deserts—areas that, with the exception of the Amazon, are relatively poor in animals and plants; the earth's richest habitats are also the places where people like to live. What's more, many of these wide-open spaces look quite different when seen in a finer grain. Measured by population density, for example, the United States has large wilderness areas in its western drylands and Rocky Mountains; look more closely, and just 2 percent of the Lower 48 states is composed of undeveloped roadless areas larger than twenty square kilometres. Drawn as a square, a space that size would take an hour to cross on foot.

Or perhaps we should look at the flagship species of wildness—the planet's most magnificent animals. In the oceans, the world's biggest fish, from tuna to cod to swordfish to sharks, have been reduced to an estimated 10 percent of their past abundance. Research over the past decade suggests that the great whales, too, may number one-tenth or less of their historical peak; even grey whales on the Pacific coast of North America, which at twenty-five thousand animals had been considered an almost fully recovered population, are now known to have a richness of genetic diversity that suggests they numbered three to five times more in the past. Imagine: seventy-five thousand whales making the yearly migration from their breeding grounds off Mexico to the food-rich waters of Alaska. On land, meanwhile, just 20 percent of the globe still houses all of the major mammal species that it did in the year 1500. That figure is twice as high as 10 percent, but with rare exceptions, those species—the 250 largest fur-bearing animals, from polar bears to elephants, from kangaroos to jaguars—are even more vastly reduced in raw numbers than in range.

To pay attention to the web of life on earth today is to acknowledge that our times are grim almost without relent. The best available evidence suggests that we exist in the accelerating freefall of what has been branded "the sixth extinction"—a fading-to-black of species worldwide at a rate that recalls five earlier spasms of mass loss imprinted in the fossil record. These range over time from the Ordovician extinction, 440 million years ago, in which 85 percent of known animal species died off, most likely through the fluctuations of an extreme ice age, to the most recent Cretaceous extinction, which sidelined 75 percent of species, among them the dinosaurs, probably in the aftermath of an asteroid's collision with the earth or a period of spectacular volcanic eruptions. Today, worst-case scenarios count as many as 36 percent of the planet's life forms as vulnerable to near-term extinction. It is not an empty threat: among species believed to have gone extinct since the year 2000 are the Chinese paddlefish, a European mountain goat called the Pyrenean ibex, and a tiny vesper bat with the lovely name of Christmas Island pipistrelle.

The news that one or another plant or animal is at the brink of eradication is now so commonplace that jaded newspaper reporters talk of "species-of-the-week" stories. Yet living things are defined not by frailty, but diehard tenacity. Three billion years ago life on earth came into being, a miracle still beyond our understanding, and has carried on through every conceivable calamity. If anything can be said to be the mysterious "spark of life," it is this singular impulse: *to be*. Extinction is not mere death; it is the death of the cycle of life and death.

The pioneering book on the sixth extinction is *The Sinking Ark*, published by the biologist Norman Myers in 1979. Myers applied a more accurate term, however—"the great dying." Extinction may hold centre stage in the global conversation about disappearance, but it is far from the whole story. Much more common in daily life is extirpation, sometimes called "local extinction"—the vanishing of particular species from particular places. Tigers are not extinct, for example, but they have disappeared from 93 percent of their original range, including the nations of Afghanistan, Iran, Kazakhstan, Kyrgyzstan, Pakistan, Singapore, Tajikistan, Turkey, Turkmenistan and Uzbekistan, as well as most (and possibly all) of China's provinces and several Indonesian islands. The Pashford pot beetle, on the other hand, is thought to be extinct, but has never been known to live beyond certain bogs in east-central England. Which is the greater loss, the extinction of the Pashford pot beetle from its few swamps, or the extirpation of the tiger from an area sixty-five times the size of the entire United Kingdom? That is a question to be weighed on the cosmic scales. Extinction wipes out, point by point, the clues to the code of existence; extirpation is the great, sucking retreat of the tide of life.

The great dying can also be a matter of simple, numerical subtraction. Consider the wolf. Once the planet's most widespread carnivore, the wolf is still present in 65 percent of its primeval range worldwide. That sounds hopeful, almost impressive, except that the animal is categorized as "fully viable" in just five of the sixty-three countries where packs once prowled; the "viable" nations include the United States, where wolves have vanished from more than 90 percent of their former range outside of Alaska. Genetic research suggests that the wolf population in North America could be as low as 5 percent of historical numbers, and declines have been even more severe in the rest of the world.

More resilient life forms appear to challenge the idea of a 10 Percent World. Behold the birds, with their tremendous adaptability and mobility. The global population of wild birds has dropped by an estimated one-fifth to one-third in the past five hundred years; from an avian perspective, ours is at worst a 66 percent world. And yet: What weight do we give to the 154 recorded bird extinctions over that same period, from the dodo on the island of Mauritius to the recently vanished po'o-uli of Hawaii? How do we account for the 190 additional bird species that are classified as being at extremely high risk of extinction, which amount to 2 percent of all our feathered friends? Where in the miserable calculus do we count the steady replacement of bird diversity by that handful of hardy, aggressive, opportunistic species—crows and starlings and pigeons and mynas—that thrive in the new wilderness of human civilization?

Other branches of the tree of life make the vision of even a 10 Percent World seem like rose-tinted optimism. Recent reviews of diadromous fish—those species that divide their lives between fresh and salt water, such as salmon, herring, eels, whitefish and sturgeon—found them to be among the hardest-hit creatures on earth. Researchers from two New York universities studied twenty-four species along the Atlantic coasts of North America and Europe; they found that every one of them has decreased by more than 90 percent from its historical abundance and more than half have declined by more than 98 percent. One species, the houting, has been extinct since 1940. I for one had never heard of it before.

Reach at random into the ecological grab bag, and one loss invariably leads you to others. When Columbus first encountered North America, it was home to as many as *five billion* black-tailed prairie dogs, rodents that live in underground colonies. Today, they endure on just 2 percent of their historical range. Prairie dogs are considered an indicator species of the health of North America's plains, and, sure enough, just 10 percent of the continent's native grasslands are still natural in any meaningful way. Most have been converted for agriculture, with the soil itself exhausted: one-third of the planet's arable land is now unproductive, and on the remainder the extraction of nutrients outpaces their replenishment. Twenty percent of the earth is still covered with ancient forests, but more than 50 percent of nations no longer have any old-growth forests at all, and fully half of the forest cover that has vanished in the last ten thousand years worldwide has been lost in just the past century.

You can take the cosmic view: the current rate of species extinction is thought to be as much as one thousand times higher than the background rate through evolutionary time. You can look at the world through a magnifying glass: in the United Kingdom, a world leader in insect conservation, scientists report that so many insect species are in danger of extinction that it would be impossible to develop recovery plans for each one individually. Even plankton, the tiny life forms that are the basic building blocks of the ocean food chain, have recently been estimated to be declining by 1 percent per year. But how long have they been in decline?

Even where there is good news, it often supports the idea of a deep biological impoverishment. In the oceans, the cardinal species of salt water—from coastal seabirds to whales to salmon to sea turtles to sharks to cod—have in recent times recovered to about 16 percent of their historical numbers, up from an estimated low of 11 percent. Yet even this pale revival is led by only the few species that have benefited from conservation campaigns; many others continue to fade.

Meanwhile, the older pattern of losses continues. Technology has made it newly possible to catch such fish species as the round-nose grenadier, which is usually found in the nightmare black-ness between 180 metres and an incredible 2,600 metres beneath the waves. Roundnose weren't commercially fished until the 1970s, but have since declined by more than 99 percent, with other deep-sea species suffering similar decreases. In fact, there is a theory that predicts these sorts of outcomes, known as the "factor of ten hypothesis" and first put forward by the marine biologist Ransom Myers. Myers proposed that human exploita-tion of a wild species tends to result in a decline to a fraction of the original population, at which point the creature in question is scarce enough that it makes less and less economic sense to pursue it. That tipping point, he said, is 10 percent.

♦ ♦ ♦

At the turn of this new millennium, Peter Kahn, the psychologist who first identified environmental amnesia among inner-city children in Houston, Texas, continued his research in Lisbon, the capital city of Portugal. He would later remember one conversa-tion in particular. "I heard," an eighteen-year-old told him on the banks of the Tagus River, "that some time ago, when there was none of that pollution, the river, according to what I heard, was pretty, there were dolphins and all swimming in it. I think it should have been pretty to see. Anyone would like to see it."

Anyone would like to see it. Counting up the ways we have wounded the earth quickly starts to feel like stacking skulls in a crypt, but the history of nature is not always and only a lament. It is also an invitation to envision another world.

Picture a map, creased and yellowed, the paper as soft as

cotton. Sketches and notations cover nearly every inch—a thousand wanderers' memories of nature as it was. There are lions in the south of France, for example. Lions in Egypt, in Israel and Palestine; lions, for that matter, in a continuous belt all the way across southern Asia. The man often called the "father of history," Herodotus, wrote of lions devouring the pack camels of the Persian army by night as it raided Greece in 480 BC. Not one of these places is home to wild lions today.

These are not sights from some ancient age of fire and ice; we are talking about things seen by human eyes, recalled in human memory, but in places where they are unimaginable today. Wolves in England and Japan and running in packs through the streets of a rising Paris. Elephants in China, brown bears in the hills of North Africa, California condors in Canadian skies, a thousand miles from the nearest condors today. Sailors report walruses at the mouth of the Thames River south of London, and in the Gulf of Maine. In the year 1377, a poet in what is now Germany speaks of entering the Great Wilderness, an unbroken forest that took three days to cross and was home to bison, wild boar, wild horses, wolves, bears, lynxes and wolverines: "Pleasantry and laughing had become hushed," he writes. In 1885, an artist draws the Hamburg Harbour sturgeon market, twenty of the huge fish laid out on the killing floor, each as large as the barrel-chested human butchers. Three thousand sturgeon a year were then being pulled from the Elbe River, where a child growing up today will never see one, may never even hear that the water was once home to these giants.

Let's not forget the sounds. On the islands of New Zealand, a ship's naturalist, a man who has sailed the far corners of the globe, is stirred awake at a distance of a quarter mile by forest birds, "the most melodious wild musick I have ever heard." Five hundred years earlier, the music was wilder still; by the time European explorers reached New Zealand, the Maori culture had long ago driven the seabirds to offshore islands and wiped out half of New Zealand's bird species, including the flightless birds known as moas, some species of which were among the largest birds ever to have lived and are known from studies of their tracheal bones to have sung deep, booming songs. One of the first generation of scuba divers, meanwhile, remembers his first dives off the coast of La Jolla, California, in the 1960s, when the spawning chant of huge white sea bass rising up from the depths was as loud as a freight train. A seventeenth-century sea captain sits in his quarters; by the light of an oil lamp he writes that sailors who lose their course in the Caribbean Sea can recover it, day or night, navigating by the noise of endless shoals of sea turtles making their way to their nesting beaches on the Cayman Islands. He does not say what it sounds like. Splashing? Breathing? Waves breaking across turtle shells?

The sheer abundance of life recorded in documents ranging from naturalists' journals to fisheries reports is an astonishment. In the North Atlantic, a school of cod stalls a tall ship in midocean; off Sydney, Australia, a ship's captain sails from noon until sunset through pods of sperm whales as far as the eye can see; in the American South, alligators dwell in "every swamp, river and lake." More than two hundred species of bird and nearly eighty of fish inhabit Manhattan. When the shad pulse up the Hudson River to spawn, they push a wave like a tidal bore in front of them, while across the continent, Pacific pioneers complain to the authorities that splashing salmon threaten to swamp their canoes. Every major land mass is criss-crossed by the migration routes of animals—ten thousand teeming journeys, many of them as breathtaking as the famous herds that can be witnessed on the Serengeti plains of Africa today. There are birds like you simply would not believe: loons "on almost every lake and moderate-sized pond" from Virginia to the High Arctic; brant geese "by the thousands and millions" that would "rise as one bird and literally darken the whole western sky"; little shore-birds that flocked up the coasts "like smoke rising from forest fires burning from horizon to horizon."

The stories go on—here is the author John Steinbeck in the Gulf of California in 1940, when "the swordfish in great numbers jumped and played about us"; here is Chesapeake Bay on the U.S. Eastern Seaboard when hunters could harvest 100,000 diamondback terrapins a year for the famous turtle soups of Baltimore and Philadelphia; here are pearl oyster reefs as tall as a five-storey building and more than ten kilometres long, creating a form of seascape that is completely unknown today.

But let me add just one personal account to this miscellany. Once, in Argentina, I saw a city disappear. I was visiting an expatriate brother, and we had grown tired of the seething core of Buenos Aires. We made our way to a park along the River de la Plata, and had just started down the crest of a dike beneath an arcade of trees when two fork-tailed flycatchers scissored the air in front of us. Small and shaded in white, black and grey, they would be unremarkable birds if it weren't for their two tail feathers, which are long and endlessly alive to the breeze. They're the kind of birds you'd expect to find in paradise.

I know people who can't see a bridge without also seeing in their mind's eye the blueprint of its structure, or who imagine what every stranger looks like beneath his or her clothes. In that moment in Argentina, I had my own perception of hidden things. The city's smog and high-rises seemed to fade, and what remained were the flood plains of the silver river, its reedy oxbows, its wooded islands, every inch alive with birds and insects and unseen, bustling beasts. This was the *understory* of Buenos Aires—the place that lived before the living city. But the vision couldn't hold. The flycatchers moved along; in the distance, the car horns resumed their endless honking.

When I returned home, Buenos Aires wouldn't leave me. In that strange way that awareness transforms into coincidence, a title later caught my eye in the rare books section of a library: *An Account of a Voyage Up the River de la Plata, and Thence Overland to Peru.* The author was the French traveller Acarete du Biscay, who disembarked in Argentina in the late 1650s, more than three centuries ahead of my own visit. In his "Description of Buenos Ayres," du Biscay opens with the frogsong that followed the rain in sultry weather. The natural world had not yet been cleaved from the city, and the Frenchman spoke of it in tones of astonishment:

> The River is full of Fish . . . there are abundance of those whales call'd Gibars,* and Sea-dogs who commonly bring forth their young ashore, and whose Skin is fit for several uses . . . there are likewise a great many Otters, with whose Skins the Savages Cloath themselves . . . most of the little Islands that lie all along the River, and Shore sides are cover'd with Woods full of Wild Boars . . . there are likewise a great many Stags.

* Apparently humpback or fin whales.

The landscape that du Biscay witnessed could hardly be considered untouched. As he planted his feet in the mud streets, some two dozen Dutch tall ships swung at anchor offshore, waiting to fill their holds, while nearly five thousand residents worked to clear grasslands for crops, cut forests for wood and hunt the vast plains known as the Pampas for game (including the one-pound eggs of the rhea, a flightless bird as tall as a man). The European conquest of the Americas was well underway, but the human touch was nothing new. Thousands of indigenous people, the Querandí, had lived in the area for millennia, though at the time of du Biscay's visit their numbers were falling to the holocaust of diseases introduced from Europe. The Querandí, meanwhile, were only one of dozens of tribes living in succession up the river system to the headwaters 2,500 kilometres to the north. The earliest Spanish visitors, too, had left their imprint, abandoning livestock that went on to invade the Pampas, changing the grasslands before later settlers could even set eyes on them. What the Pampas looked and smelled and sounded like before they were grazed by feral cattle and horses was not recorded. It will never be known.

Still, du Biscay was in awe of the natural wealth of Argentina, and in return the people of Buenos Aires told him a story. From time to time, they said, the settlement was threatened by buccaneers or foreign armadas, and when these enemy ships appeared on the horizon, the men would mount their horses and haze the grasslands. They would herd together the free-ranging bulls and cows, the mules, asses and horses, and also the native animals, the llama-like guanacos, the deer, the vicuñas with their wondrous wool. Then they would drive all of these animals thundering toward the shore. Picture the scene, the air shuddering, dust ballooning, every living thing without a hole to hide in driven forward as if chased by a brush fire— tortoises, snakes, lizards, voles and mice, armadillos, foxes, wild cats, ground birds, songbirds, even vultures, even locusts. All would crush to the river's edge, there to seethe and buck and blow, a storm lit from within by the flash of teeth and the whites of eyes, and in that moment the strategy was complete:

> 'Tis utterly impossible for any number of Men, even tho' they should not dread the fury of those Wild Creatures, to make their way through so great a drove of Beasts.

A wild abundance so overwhelming that it could be used as a military defence—this is one way of picturing the living world of the past. Today, the *Environmental Atlas of Buenos Aires* lists no creature larger than the rhea, which is found only "in captivity or partial freedom"; the maguari stork, which "appears occasionally"; and the capybara, the world's largest rodent, listed as "receding."

In the end, it didn't prove impossible. We did indeed make our way through so great a drove of beasts.

Chapter 4.

THE OPPOSITE OF
APOCALYPSE

———— ·◆◆◆· ————

People often say that only a disaster will convince us to change our ruinous relationship with the natural world, or that our grandchildren will curse us for the damaged planet that we'll leave them. The history of nature suggests that neither statement is likely to prove true. Our own ancestors handed down a degraded globe, and we accepted that inheritance as the normal state of things. As our parents and grandparents did before us, we go about our lives in the midst of an ecological catastrophe that is well underway.

Our baselines have shifted. But it's one thing to recognize that amnesia, and another to say what the original baseline actually is. Where in the billions of years of life on earth could we possibly draw that line?

In the Americas, there is a traditional answer. In fact, the baseline can be nailed down not only to the day, but almost to the hour: about 2 a.m. on October 12, 1492. That's when a sailor named Rodrigo de Triana, under the command of Christopher Columbus, spotted what would come to be known as the New World. Beginning with that moment, nothing would ever be the same.

It's still widely believed that the European explorers made landfall on a wild landscape, in which the Native Americans were a part of the natural balance or at worst a minor disturber of its harmonies. That view has helped shape the Americas—and, by extension, the world. In 1963, for example, the zoologist Starker Leopold authored a committee report that became a grounding philosophy for America's globally influential national parks system. "As a primary goal," he wrote, "we would recommend that the biotic associations within each park be maintained, or where necessary recreated, as nearly as possible in the condition that prevailed when the area was first visited by the white man. A national park should represent a vignette of primitive America."

In the public imagination, the original state of nature in Australia, the South Pacific, and much of Africa and Asia has also been widely understood to be the world as it was witnessed by the first European explorers and conquerors. A different standard has been applied to Europe itself, and to other regions—such as China and the Middle East—that are popularly recognized as having had advanced societies that transformed their landscapes earlier in history. As Europeans grow more interested in restoring some vestige of pristine nature to their continent, however, the question soon arises: Restore to what? Looking for a turning point at which hunting, land clearing and fishing suddenly accelerated, many draw a line at the dawn of the Roman Empire. Russia's *ʒapovednik* (zah-po-VYED-nik) protected areas—the oldest of which were created not long after the first U.S. national park—were established with the intention of maintaining *etalon*, a word usually translated as "baselines," and any human use other than research is technically forbidden.* It is an unmistakably biblical view of natural history: before civilization, there was Eden.

North America remains the world's most iconic fallen garden, having gone from an apparently endless wilderness to a settled super-civilization in just five hundred years. Perhaps the greatest symbol of the price of that transition is the passenger pigeon. The passenger pigeon was a bird of eastern North America, quite similar to a modern-day city-park pigeon, except that where those urban birds shimmer grey, purple and green, passenger pigeons glittered blue-steel and rust. In the early 1800s, they were thought to be among the most abundant birds on earth. The ecologist Aldo Leopold eulogized them as a "living wind," John Muir recalled their "low buzzing wing-roar," and every town in the eastern U.S. seems to have seen them blot out the sun or set down in such numbers that they snapped the limbs off oak trees. It's often forgotten that they lived in Canada, too, but Toronto knew them in flightlines one hundred yards deep and stretching as far as the eye could see. That city's neighbourhood of Mimico is said to take its name from the Ojibway word *omiimiikaa*, meaning a gathering place for passenger pigeons.

---

* In reality, many of these protected areas have been hard hit by commercial and industrial activities, both sanctioned and unsanctioned.

Unlike the bison, the pigeon did not survive the colonial onslaught. The last known passenger pigeon, named Martha, died in the Cincinnati Zoo on September 1, 1914. The species is extinct. The passenger pigeon has lived on only as a transgenerational memory of the astonishing natural plenitude that existed ahead of the colonial era, and the ultimate cost of human greed.

Today, the symbolic importance of the passenger pigeon is being questioned. In his book *1491*, author Charles C. Mann presents evidence that the awe-inspiring flocks of passenger pigeons remembered from the nineteenth century were not natural, but rather "pathological"—symptoms of an ecological system thrown wildly out of balance. Others, such as University of Utah wildlife biologist Charles Kay, argue that all the familiar examples of wild abundance taken from history—the thundering buffalo herds, the immense salmon runs, the seemingly endless supply of beavers—were nothing more than unnatural population outbreaks that took place when those animals were suddenly freed from hunting pressure and competition for food. The missing hunter and competitor was none other than *Homo sapiens*, human beings—in this case, North America's indigenous people. Lacking immunity to foreign diseases such as smallpox and scarlet fever that were introduced by European explorers, the native people of the New World died in epidemics that may eventually have claimed 90 percent or more of their number; in many cases, the diseases spread from tribe to tribe so quickly that whole cultures were devastated before they had encountered a single pioneer. By the time European settlement began in earnest, the natural world was going wild on a landscape that was more empty of human influence than it had been for millennia. Before that, according to Kay, "Every inch of North America was artificial." He argues that the normal state of nature before Columbus was a landscape on which people had hunted wildlife into scarcity, a condition he refers to as "aboriginal overkill."

Anthropologists and archaeologists have known for decades that the indigenous people of North and South America were not mainly scattered hunter-gatherers, but instead had complex, advanced and often densely populated societies. That idea has been slow to find mainstream acceptance. The founding stories of Canada and the United States have long rested on the justification that European explorers settled a wild and thinly populated continent. Surviving indigenous people, on the other hand, reject the suggestion that they have no real claim to the territories that historically sustained them, though they also benefit from the myth that they once lived in perfect harmony with nature. To accept that native cultures had the numbers, knowledge and power to transform entire continents lays waste to the widely treasured ideal of wilderness.

The charge that the passenger pigeon is a false symbol of past abundance is not new; observers at least as far back as the 1800s note that the earliest American colonists made no mention of pigeon flocks as spectacular as those witnessed later. The birds' bones also do not turn up in great numbers in the prehistoric garbage heaps known as middens, considered to be among the best available records of what people ate ahead of European arrival. "The simplest explanation for the lack of passenger pigeon bones," comments one midden-searching archaeologist, "is a lack of passenger pigeons. Prior to 1492, this was a rare species."

The simplest explanation for any phenomenon, though, is often an oversimplification. Enrique Bucher, an Argentine expert in colony-forming birds, reviewed the competing theories about how the passenger pigeon went extinct. He concluded that it was impossible for the pigeons to have been rare in indigenous North America, because sheer abundance was the birds' survival strategy. Their primary foods were the beechnuts, acorns and chestnuts that are collectively known as "mast." Mast forests are curiously unpredictable, often producing nuts in a boom and bust pattern that shifts over time. The passenger pigeon had adapted by forming flocks of millions that could line out over huge areas of countryside in search of the trees that were bearing that year's crop. A special call alerted other birds to the loaded trees, and the famously dense roosts were meeting places from which hungry birds could follow others that had found food. In other words, passenger pigeons had the paradoxical ability to be both numerous and uncommon: The huge flocks were only occasional visitors to any one place, and years could pass before they returned. They were only perceived to be widespread after European settlers themselves had dispersed widely.*

It's certainly possible that the nineteenth-century flocks grew even larger as they benefited from a lack of competition for mast, which had been a staple food among indigenous nations. But the flocks in 1491 would still have been mind-boggling to a time-traveller from today; at the least, they were probably comparable to the passenger pigeon's little-known living relative, the eared dove. In South America, these doves continue to

---

* Taste is another reason that passenger pigeons may not have turned up in bone heaps until historical times—some Native American commentators compared the flavour of pigeon meat to skunk, and it was only after other game was scarce that it became a preferred food among European colonists.

form colonies of over ten million birds, and for decades have both competed with humans for their crops and been the target of eradication efforts far more unrelenting than anything indigenous communities could have mustered, including poisoning, trapping and unregulated hunting—all without any lasting effect on dove populations. Only recently have the birds begun to decline, apparently due to the destruction of their grassland habitat. Passenger pigeons probably faded for the same reason. By the 1860s, four-fifths of the forest they relied on had been levelled. Suddenly, no amount of searching could turn up sufficient mast, and the pigeon population—still heavily hunted—went into freefall. They were such highly social animals that the last scattered flocks and breeding pairs may have starved to death even in the presence of food, simply from the stress of isolation. One of the last passenger pigeons seen in the wild was among a flock of domestic pigeons.

The belief that North America before Columbus was an unspoiled Eden is unfounded, but there is no reason to declare that the continent's most natural condition is one in which most species have been hunted into scarcity. There clearly were densely populated areas, such as the Mississippi Valley, where the first European visitor, Hernando de Soto, in around 1540 reported many Native American cities but not a single buffalo. But there were also those places less hospitable to human presence. The first European to cross the Texas Panhandle, Francisco Vásquez de Coronado, in 1541, witnessed uncountable bison and was never once out of sight of the animals. The anthropologists Torben C. Rick and Jon M. Erlandson surveyed the evidence that ancient peoples depleted and otherwise altered coastal areas from Alaska to the Gulf of Maine to California's Channel

Islands, and found the proof was "overwhelming" that they "often" did. The two researchers also speculate, however, that the struggle to survive in increasingly degraded surroundings may have given rise to the conservation values that many indigenous nations gave voice to at the time of European contact. If so, then 1492 was a clash of shifting baselines: European nations revelling in the discovery of apparently infinite natural riches just as the Americas' original cultures were formulating a widespread understanding of ecological limits.

♦ ♦ ♦

There have always been corners of the globe where the human influence fades and a more ancient order asserts itself. In some cases, these are simply places too high, too dry, too cold or too barren for long-term human survival. But there are many other examples. Throughout history, large areas have been set aside as the hunting grounds or pleasure gardens of kings, queens and other social elites, preserving the trees, plants and wildlife against the ordinary onslaught of settlement—at one point, royal "forests and chases" covered one-fifth of England. At the turn of the year 1000 BC, Europe was home to forests in which every human activity but the worship of gods was forbidden. In western North America, no-man's-lands between hostile indigenous nations were far richer in game than zones where people hunted freely. Today, of course, we have parks and protected areas, but more particular traditions continue as well. Fly over the small African country of Malawi, for instance, and you see a landscape dotted with the remaining groves of an otherwise vanished forest; these miniature jungles are the graveyards of Chewa villages and the sacred ground of all-male secret societies called upon to communicate with the dead.

Wilderness has long been understood as the original, wholly natural condition of any landscape. It has been celebrated by writers, artists and philosophers, and inspired generations of environmental activists. Yet while popular wisdom holds that every place on earth has a single, wild condition that it will maintain unless altered by human hands, science has moved in the opposite direction. Nature has proved harder to pin down than expected.

The turning can be traced to a competition of ideas in the early twentieth century, beginning with the American botanist Frederic Clements. Clements grew up on the tallgrass prairie of Nebraska just as the stands of bluestem and switchgrass, up to eight feet high and with roots nearly as deep, were making way for tilled fields. The last buffalo herds had vanished while he was a boy, though he would have heard them remembered, or even walked the animals' old migration routes, some tamped so hard by beating hoofs that their imprint is still visible from the air today. He would have known pronghorn antelope, prairie dogs, rattlesnakes and the rasping call of the sandhill crane, so raw and unformed to our ears that it seems to cry out from pre-history. He would have known all of these, and the loss of them.

In 1916 Clements published *Plant Succession*, one of history's most influential books of ecological ideas. In it, Clements proposes that the mature condition of every landscape is fixed—predetermined by soil conditions, climate, rainfall and other limiting factors. Disturbances could set the process back, but the land would then recover through a predictable series of stages. Nature was always on a path to perfection.

While Clements's pure science was little known outside of scholarly circles, his basic idea took hold in the popular imagination. With the Dust Bowl in the 1930s, when the overworked soils of America's drought-struck plains were sucked up by the wind to dirty the sails of yachts off the Atlantic seaboard, it seemed clear that a timeless and durable natural order had been disrupted. Clements even recommended a place to look for the original prairie: in cemeteries, one of the few places where native plants and grasses had not been completely ploughed under.*

The challenge to Clements's view arose when researchers went looking for landscapes that had reached their ultimate climax states. Another American botanist, Henry Gleason at the University of Michigan, observed in the 1920s that while plant communities generally do become more complex over time, they rarely if ever do so in the same way. Oxford ecologist Arthur Tansley, meanwhile, found that many different combinations of species appear on landscapes that otherwise share the same soils and climates. Tansley also pointed out that in Britain and much of Europe, stable, fully mature ecosystems had developed that were entirely the result of human influence—climax states of disturbance.

It was decades before these counter-arguments overthrew Clements's theories. Today, the Clementsian idea that every place on earth has a single, predictable state of nature that it will eventually achieve if left undisturbed has been discarded. This would appear to spell doom for the notion of a natural baseline, but in fact: not so much. Nature may be a more complex

---

* This is famously echoed in Aldo Leopold's 1949 book *A Sand County Almanac*, in which he refers to a Sauk county cemetery as a "yard-square relic of original Wisconsin."

blend of chaos and order than Clements believed, but that doesn't mean that human impacts are just another signal lost in the noise. By human-scale timelines, the ordinary rate of change on many wild landscapes is so slow, patchy and incremental as to be imperceptible without dedicated study. To say that change occurs in nature is not the same as saying that every change that occurs in nature is equal, a fact that we recognize in our most day-to-day observations: a farm is not a forest, after all, and a parking lot is not a farm. There is an ever more urgent need, as the environmental historian Donald Worster puts it, to hold fresh "the memory of a world by which civilization could be measured." Research today considers new questions: How much balance is normal in nature, and how much change? At what point does change become damage? In what ways is natural change different from the changes wrought by human influence?

One place to look for answers is in "macro time," or time that is measured in millions of years and renders the whole of human history to the blink of an eye. We know that the earliest signs of life on this planet date back 3.5 billion years. But is there a point along the timeline where we can say that this thing we call "nature" began?

In fact, there is. The story begins in the late Permian period, approximately 250 million years ago. At a glance, nothing Permian would seem recognizable today. The world was hotter and drier and had a single supercontinent known as Pangaea, which was probably home to the world's first warm-blooded vertebrates. What passed for big, charismatic wildlife was typified by the gorgonopsids, vaguely dog-like creatures that could stand as tall as a modern-day bear but with much more impressive teeth. Among their prey were the cynodonts, which were something like oversized, prehistoric badgers. Life since the Permian has never been easy—the period ended with a poorly understood mass extinction that erased more than 90 percent of marine species and 70 percent of creatures on land. But from the Permian forward, nature has always re-emerged in a recognizably similar form, with warm- and cold-blooded animals; a mix of carnivores, herbivores and omnivores; and a steady tilt toward diversity of life.

At least five major changes have taken place in that multi-million-year-old pattern since our own species began its spread around the globe. For one, we have largely eliminated the mega-fauna of our age. For at least 250 million years and through every variation in global climate, life on earth was not so empty of large and fierce beasts as it is today except during periods of cataclysmic mass extinction. The same is true of "keystone species," or the animals and plants with the greatest influence over the structure of their environments; no ordinary force in deep time has ever tended to erase them in the way that humans have. Until the human era, other species also enjoyed the freedom to move, unbound by the fences, highways, cities and other barriers that restrict them today. Nonetheless, they tended to arrive in new places at a pace that permitted gradual adaptation. There was no possibility, for example, of a species like the rabbit appearing suddenly as a breeding population on a new continent, as occurred several times when they were introduced by Australian colonists; with no natural predators, the rabbit population exploded in the mid-1800s, causing such devastation through overgrazing that it was among the leading contributors to native animal and plant extinctions on that continent.* Lastly, we now live in a time in which the actions of a single species—us—appear to be dramatically changing the climate. While not unheard of in prehistory (luckily for us, single-celled algae helped bring about an oxygenated atmosphere 800 million years ago), it is fair to say this is extremely rare.

There is, in other words, a fundamental difference between background patterns of change and cataclysmic change—and we, human beings, belong in the cataclysmic camp. Us and killer asteroids. Us and glaciers two miles deep and as wide as continents. If nature's complexity has been described as "noisy clockwork," then the singular tone of the human era has been that of a pendulum slowly winding down. Taken together, these alterations to the patterns of nature are less the "environmental challenges" that we speak of today than they are breaches in the space-time continuum.

◆ ◆ ◆

In September 2004, fourteen leading conservation thinkers—described by an organizer of the group as "National Academy, silverback, rock-star scientists"—gathered in a century-old stone-and-mortar house on the Ladder Ranch, a property in New Mexico belonging to media mogul and philanthropist Ted Turner. The goal was to draw a radical new baseline for the way we look at nature.

---

* Rabbits are an introduced species in many parts of the world, though they often did not immediately succeed. After their initial introduction to England in the 12th century, they fared poorly and struggled to burrow in the dense soils. A special tool was designed to help "warreners" dig the rabbits' burrows for them, or they were offered "pillow mounds" of loosened earth in which to dig themselves.

The group had been influenced by the concept of "rewilding." The term was coined in the early 1990s by Dave Foreman, a founder of the direct-action Earth First! environmentalist network, writing in *Wild Earth* magazine. Foreman used rewilding in the general sense that the word evokes: "to make a place wild again." In 1998, conservation biologists Michael Soulé and Reed Noss applied the term to science, defining rewilding as the effort to restore wilderness on a large scale. Once again, the familiar question arises: Restore to what?

The Ladder Ranch workshop's answer appeared the following year in the science journal *Nature*. The baseline state of nature for any given place on earth, they declared, should not be drawn in 1492 or even the dawn of civilization. Instead, it should be drawn at that point in time when the most complete web of life existed during the current climate cycle in geological time. That would be the end of the last ice age, also known as the Pleistocene epoch, when the glaciers receded to reveal the landforms that we know today, while the planet warmed to produce the climate that has endured ever since.

The last continent on earth to be populated at the end of the Pleistocene was North America, some fifteen thousand years ago. At that time, the land was home to a hair-raising assortment of megafauna. Most of us are familiar with some of these giant animals, like the mammoths and mastodons that look like woolly elephants and, like elephants in some parts of Africa today, may have roamed grasslands in densities of more than three animals per square kilometre. But North America was also home to pampatheres, which resembled armadillos the size of overturned rowboats, and another armoured family, the glyptodonts, at their largest the size of a subcompact car. There were ground sloths—amiable-looking herbivores that often stood up on their hind legs and could weigh nearly three tonnes. There were herds of wild horses, some as heavy as today's Clydesdales, and tapirs rooting through the wetlands, and an antelope called the saiga, with a pouchy snout that acted as a dust filter.* Wild oxen drank at waterholes alongside camels that would tower over today's dromedaries. There were giant moose, giant llamas, giant elk, giant boars. There was a beaver nearly the size of a bear.

There were monsters. Packs of dire wolves were widespread, the animals outweighing modern wolves by twenty pounds apiece but still far from the most fearsome predators in the Pleistocene wilderness. That title probably goes to the short-faced bear, a flesh-eater large enough to look you in the eyes while still on all fours. The greatest feline, still haunting the pop-up-book nightmares of children, was the sabre-toothed cat, with serrated fangs as long as chef's knives and a body twice the heft of a modern lion's. There were lions, too—prides of American lions that were like today's African and Indian lions in every way but one: they were larger.

The list goes on, from scavenging birds with wingspans of fifteen feet to beetles adapted to rolling the dung of giant animals. Then, North America's megafauna began to disappear. Scientists have debated the cause of the mass extinction for decades, but evidence increasingly points to the spread of humans around the globe at a time of intense climate change. Go to any corner of the planet, and the moment that *Homo sapiens* first shows up in that place will be roughly the time that many of its largest species begin to fall toward the void

* The saiga still exists as a critically endangered species in Central Asia.

of extinction. Africa is the exception, where megafauna such as elephants, giraffes, lions and hippopotamuses evolved alongside people. Otherwise, the pattern holds. Fifty thousand years ago, humans reach Australia and twenty-one entire genera (groupings of species with similar characteristics) disappear over the following millennia; every land-based species with an average weight above one hundred kilograms is wiped out. Thirty thousand years ago, modern humans settle in Europe, and nine genera vanish. North America loses thirty-three, with the Americas as a whole shedding 75 percent of their big beasts. Most compelling is the fact that the die-offs taking place on continental mainlands were often postponed on offshore islands, for the simple reason that humans had not arrived on them yet. The biggest animals in the Caribbean, including sloths and various oversized rodents, survived the extinctions going on in North and South America only to go extinct themselves when our species reached the islands just six thousand years ago. Despite being Australia's neighbour, New Zealand didn't lose its large fauna—eleven species of flightless moa—until just eight hundred years ago, with the arrival of the Maori culture.

What the Ladder Ranch group proposed was an attempt to revisit that older world—"Pleistocene rewilding," they called it. A series of experiments could be undertaken, they wrote, in which existing species from one part of the planet—say, elephants—might be carefully "reintroduced" to places where similar species—like mastodons—went extinct in the Pleistocene. The end point, described as "the ultimate in Pleistocene rewilding" for North America, would be a free-living population of lions somewhere on the Great Plains, limited only by perimeter fencing.

Without megafauna, the scientists argued, the planet's landscapes would forever be ecologically incomplete.

Response to the idea was explosive. Opposing scientists pointed out that the rewilding article appeared in the same issue of *Nature* as a study showing that lion attacks on humans in Tanzania had risen 300 percent in fifteen years—a testament to the risks of living alongside megafauna. *Good Morning America* broadcast a brief interview with Pleistocene rewilding's proponents, followed by a clip from the film *Jumanji* of elephants crushing cars. Several threatening letters were turned over to police, while surprise offers of rewilding habitat rolled in from ranch owners in Texas, Arizona, Kansas. Then Hurricane Katrina began to gyre toward the U.S. coast, and the Pleistocene was soon forgotten amid the horrors of that storm's landfall in New Orleans. The excitement over rewilding had lasted less than a month, but was nonetheless unprecedented; one supporter called it "the largest ecological history lesson in American history."

♦ ♦ ♦

The meaning of "rewilding" has continued to evolve, and its most widespread definition today is simply "to make wilder." The term is increasingly used to acknowledge that it may not be possible to achieve some specific natural condition from the past, the way a team of heritage experts might restore an old cathedral, and instead describes attempts to bring back species and ecological processes that have been shunted aside—to give nature fuller expression in a world in which it is muted. Under this banner, rewilding could refer to the campaign to release endangered California condors in Oregon, an effort that involves the restoration not only of the birds but of the cultural memory that condors are not creatures of the desert—the only place they are found today—but once also soared through the fog-bound forests of the Pacific Northwest coast, where they fed on the carcasses of whales. In Europe, rewilding often refers to the gradual return of bears, wolves and other animals to the landscape as rural people continue to move to the cities, though the land itself remains indelibly marked by thousands of years of human presence. Perhaps the best-known example of European rewilding is the Oostvaardersplassen nature reserve in the Netherlands, where herds of wild cattle, horses and deer run free on a savannah that, were it not for dams and dikes built in the twentieth century, would be an ocean inlet; at sixty square kilometres, the reserve is tiny by modern conservation standards and so far lacks any large predators. Even efforts to clone extinct species such as the passenger pigeon and thylacine could be considered rewilding, though only one clone of a vanished animal—the Pyrenean ibex, a subspecies of wild goat that went extinct in 2000—has ever been born, and survived just a few minutes before dying of physical defects in its lungs. A member of the team working on thylacine cloning once placed the odds of success at 30 percent over the next two hundred years.

Few people are aware that a first step in the Ladder Ranch group's Pleistocene rewilding proposal is also underway. Not with a lion, no, nor even a camel, but rather a tortoise. It seems an appropriate choice. With their slow pace and self-contained habits—some tortoises spend 90 percent of their lives inside or within a few steps of their burrows—it is hard to believe that tortoises still share the modern world with you and me. Have they no interest whatsoever in living large? In running with the bulls or falling in love on a beach under unfamiliar stars? Their defensive strategy of hiding in plain sight seems to make the argument that certain species really are doomed to extinction by the imperfections of their nature. On the other hand, the tortoise as a life form has been on earth roughly 200 million years longer than the human model.

In 2006, conservation biologists transferred twenty-six bolson tortoises from a captive colony in Arizona to Armendaris Ranch, another Ted Turner property in New Mexico. The tortoises are unexpected beauties, with carapaces coloured jet black and topaz yellow. Although they remain fenced in at the ranch, the barriers will eventually be taken down and the tortoises will once again roam free in an area they haven't called home for ten thousand years. The bolson tortoise is making prehistory.

Out on the range at Armendaris, it seems hard to believe that the bolson tortoise could ever have been eradicated even from the single, sprawling valley that houses the ranch. The sere and shelterless landscape gives the immediate impression that humans never lingered here long, and in fact the desert basin is so inhospitable that it has been known since the days of the Spanish conquistadors as the Jornada del Muerto—the "journey of the dead man." Yet at their peak after the ice age, bolson tortoises likely roamed from at least Arizona to western Texas and southward to central Mexico—an area of perhaps a million square kilometres. Today, they exist only in a few desert basins, known as bolsons, north of the Chihuahuan Desert city of Torreón, Mexico.

The slow fade probably began with the very first encounter between a human being and a bolson tortoise, many millennia ago—the tortoise would have retreated into its shell, forelegs covering its eyes; the human would have carried it away for an easy meal. That pattern has repeated itself endlessly, beginning with the indigenous cultures, represented today by the Apache and Uto-Aztecan tribes that lived off the desert as hunter-gatherers and salt traders for thousands of years. Then came the Spanish, feeding their expeditionary parties, and sometimes whole armies, off the landscapes they moved through. They were followed by settlers—the shepherds and ranchers who, with the coming of cool weather each fall, manned signal fires on the highest summits to warn of Comanche and Kiowa raiding parties. When a rail crew of three hundred badly paid men began to move north from Mexico City into the desert basins, they were happy to find tortoises to roast over their fires. The expansion of railways and roads opened the desert to the Great Wax Rush of the early 1900s, when demand for waterproofing for military tents and ammunition sent thousands of *candelilleros* out to harvest the waxy candelilla plant, the men living off the land, eating jackrabbits, eating tortoises. In the 1920s, many bolsons were razed for farming, the mule-driving teams culling ten or more big tortoises for Saturday-night fiestas and saving the largest for their breakfast. Then came Mexico's Highway 49, and the oil survey roads, and again the hungry construction crews. Truckers would stop at the roadside to pick up campesinos' ox-cart-loads of tortoises, southbound for Mexico City or north to California, where "turtle" soup was in fashion.

Finally, in 1959, the same year NASA selected its first astronauts for human space-flight, biologists "discovered" the bolson tortoise—the largest terrestrial reptile in North America was almost extinct before it was even known to science. Efforts to save the species followed, and in 1977, the last wild bolson tortoises were protected in a UNESCO biosphere reserve, the core of which is so remote it is known as the Zona del Silencio, the "zone of silence." Humans continue to encroach on the area, however, and today the reserve is home to more than seventy thousand people. The tortoises turn up as pets in ranch houses or as bones in the ashes of cowboys' campfires, and every year, new and illegal roads are pushed into their habitat. The tortoise today exists in 1 percent or less of its prehistoric range. The miracle is not that they have disappeared from so much of the world they once inhabited, but rather that they have survived at all.

The bolson tortoise reintroduction at Armendaris Ranch will test a number of things. The animals' success or failure in the northern Chihuahuan Desert may help to finally answer the vexing question of whether ancient climate change or the spread of human beings led to the disappearance of the world's Pleistocene megafauna. The experiment has encouraged debate about whether to measure today's nature against the relatively recent past of recorded history, or to reach back through the millennia to account for every human impact. But stand in the presence of a bolson tortoise, and the living, breathing animal itself seems valuable enough. The beast is nothing short of a miracle, able to survive months without food or water and to live more than a hundred years. Its greatest contribution is its burrow—no other creature in the Chihuahuan Desert digs as big a hole. Bolson tortoise burrows plunge six feet into the earth and often extend to the length of a stretch limousine; inside, the temperature stays relatively steady through the extremes of summer wildfire and clear, cold nights beneath the picked-out winter stars. One study found 362 other species making use of tortoise excavations; in New Mexico, burrowing owls,* box turtles and skunks

---

* An endangered species in Canada; they very rarely dig their own holes.

promptly moved in. Rainwater feeds into the water table through tortoise holes, and the mouths of the burrows—which resemble bomb craters—are made up of nutrient-rich soil dredged up from the depths. Mounds used by tortoises for years exhibit a much greater diversity of plant life than surrounding areas. The result is a more varied landscape, with more niches to be filled by more species—a strengthening of the functional fabric of nature. The rewilding of the tortoise in its ancient habitat represents not only the species' slow drift away from extinction, but an overall movement toward a more plentiful world. What the bolson tortoise reminds us is that it is ultimately less important to choose a baseline than it is to choose a direction. The direction the tortoise points to is the opposite of apocalypse.

Fig. 2.

THE
NATURE
of
NATURE

How, then, am I to find you,
if I have no memory of you?

ST. AUGUSTINE

Chapter 5.

A BEAUTIFUL WORLD

———— · ◆ ◆ ◆ · ————

Not long ago, I sat down with my father to ask about his boy-hood on the North Atlantic shore. We were at my kitchen table; a bottle, increasingly empty, stood between us, and every few minutes we would hear the distant howl of the city's mono-rail. Dad was remembering the smelt that used to run up a creek near his family's home to a pool the locals called MacGillvary's Pond. Those were days of celebration, he said, when every kid in the district—it had the unpromising name of Low Point—would turn up with a basket or a bucket to fill with the little silver fish, and by evening smoke from the oil dripping into cooking fires would be hanging over the bay. He doubted the smelt still spawned there, and I replied that ours was an age of emptiness—"an ugly world," I called it. He looked at me then with the fierce eyes that used to turn me to stone as a child. "I don't live in an ugly world," he said. "I live in a *beautiful* world."

It's an easy truth to lose sight of. A few days later, I made my way from that same apartment to a patch of green in the heart of the city of millions where I've lived for more than a decade. A small lake is tucked away in the park, and though I had walked around it many times before, this day would be different. I had resolved to spend sixty minutes, just a single hour, giving nature my fullest attention.

The results were immediate. I'd always noticed that the pond was home to that most familiar of ducks, the mallard. Now I carefully examined every bird on the water, and was surprised to count ten other species, among them geese, mergansers, coots, cormorants and the northern shoveller with its oversized, spoon-shaped bill. It was a peaceful scene, the ducks push-pushing across the surface or dunking their heads to feed in the shallows. Then the lake exploded. Every bird was in motion, scrambling for the reeds, diving, bursting upward into flight. In the trees along the shoreline, the songbirds spiralled down into the undergrowth, and the air rang with chittering and honking and squawking. I saw the shadow first—a dark triangle rippling over the unquiet water, and then the eagle itself, sweeping downward as fast as I could lower my eyes.

A bald eagle hunting ducks. It tipped its wings uncertainly, then bore down with terrible speed on three dabblers, which desperately laved the water with their wings and piped a call that spoke to me more of dignified effort than of terror. At the last instant the raft of ducks split apart, and the eagle weaved at one, then another, and in that moment of indecision all three ducks disappeared into the reeds. The eagle glided upward, talons empty. Then it circled back the way it had come and settled in again to watch the pond from a line of trees.

Life, death, the great wheel of eternity! Here it all was, in the heart of the city. Yet one observation stood out above all others. Dozens of people surrounded the pond—women jogged, children played, men threw sticks for their dogs to retrieve. Not one of them showed the slightest sign of having noticed the drama that had just played out before their eyes. If I considered that I had walked around the water's edge perhaps a hundred times, then ninety-nine times out of those hundred I had been just as oblivious as everyone else.

◆ ◆ ◆

The crisis in the natural world is one of awareness as much as any other cause. As a global majority has moved into cities, a feedback loop is increasingly clear. In the city, we tend not to pay much attention to nature; for most of us, familiarity with corporate logos and celebrity news really is of more practical day-to-day use than a knowledge of local birds and edible wild plants.* With nature out of focus, it becomes easier to overlook its decline. Then, as the richness and abundance of other species fade from land and sea, nature as a whole becomes less interesting—making it even less likely we will pay attention to it.

It has become possible to lose sight of nature almost entirely. At least as remarkable, however, is the fact that many people

---

* A friend pointed out that the scene I witnessed at my local pond might have been more likely to attract people's interest on the internet. I found a number of online videos of eagles hunting ducks, including one with more than 150,000 views, along with a photo series featuring a passerby who fails to notice even a high-speed mid-air collision between two eagles, one of which crashes into the water. The photographer suggests that the bedraggled eagle that comes ashore "symbolized America in its current trials."

retain a hunger for nature against the odds. The author Italo Calvino made a memorable character of this impulse in 1963. His Marcovaldo, an Italian labourer who spends his days unloading boxes at the fictional Sbav and Co., possesses "an eye ill-suited to city life":

> Billboards, traffic lights, shop windows, neon signs, posters, no matter how carefully devised to catch his attention, never arrested his gaze, which might have been running over the desert sands. Instead, he would never miss a leaf yellowing on a branch, a feather trapped by a roof-tile . . .

Marcovaldo is a romantic, even a heroic character—but he is also a fool. He eats mushrooms that he finds in the city and ends up in hospital having his stomach pumped. He longs to trap one of the wild woodcock that fly over his city in autumn, but captures only a pigeon as domesticated as he is. When he tells his children about the forest that lies at the edge of town, they think he must mean the billboards that line the superhighway, and cut them down for firewood. Marcovaldo longs for the natural world, but he knows next to nothing about it. Its specific meanings have been lost to him.

When it comes to nature, our current global generation is the most blinkered of all time. History has left behind thousands of traces of a former attentiveness to the living world that can only seem alien to us now. One of the great mysteries of the early cave paintings in Europe and Australia, for example, is the way they represent details of large and dangerous animals. You might suggest that these long-ago artists extrapolated from the corpses of the animals they hunted, but this explanation isn't equal to the evidence. Cave art is practically defined by its extraordinary power to transmit the *living* energy of animals; it is hyper-real, in the same way that a watercolour painting can sometimes show a scene more truthfully than a photograph. One wall of Chauvet Cave in the Ardèche Valley, France, features thirty-thousand-year-old portraits of lions that convey the most subtle and fleeting of the animals' facial expressions: moments of contentment, apprehension, uncertainty. To witness such intimacies before the age of zoos or telephoto lenses would have demanded long hours watching the animals at close range. More than that, it would have required a human being to be just another species on the landscape, in the same way that, in moments of truce, lions and their prey are often seen resting nearly side by side.

The English language once had a word for the characteristic impression that a plant or animal offers to the eye. We called it the "jizz," and the adoption of that term as sexual slang is unfortunate, as it seems unlikely we'll come up with a replacement. It is the jizz, for example, that allows a skilled birdwatcher to know a bird by its silhouette alone, or by some quality of movement or the way it holds its head. The strangely unsteady flight of the turkey vulture, the flat forehead of the Barrow's goldeneye, the endless headlong running of sanderlings on a mud flat—each of these is the jizz. It is so pure an essence that, if captured in a few rough lines drawn with charcoal, it can express an animal more authentically than a portrait by a trained artist who has never carefully watched the creatures he paints. It's the jizz that ancient art so often represents. While looking at Egyptian treasures in a museum, I felt a rush of nostalgia when an engraving of a scarab beetle reminded me that I used to see a related species, the tumblebug, or *Canthon simplex*, roll balls of dung across my home prairie. I had completely forgotten; it took a 3,500-year-old artifact from another continent to make me remember.

We tend to associate this deep, almost innate awareness of nature with the Stone Age or with tribes of hunter-gatherers, but it isn't so distant even in Western culture. Professional hunters in Germany in the eighteenth century were expected to be able to look at a wolf's tracks and determine not only its size, sex and rate of travel, but also whether or not it was rabid. We remember England's "terms of venery"—the jargon of hunting—for giving us specific words for groups of animals, such as a school of fish or a pride of lions, and also for such quaintly forgotten phrases as "a tiding of magpies" and "a kindle of cats." Experts suggest that many of the terms that amuse us today—"an unkindness of ravens," "a shrewdness of apes," "a disworship of Scots"—were fanciful even in their own time and never in common use. The true language of venery, however, did more than describe beasts by the bunch; it richly evoked their behaviour. The lark's habit of flying into the air to sing was known as "exalting." The nocturnal song of nightingales was called "watching," from the idea of keeping a watch through the darkness. Venery's description of animal sounds was poetic, but also accurate: weasels really do "squeak," mice really do "cheep." Goldfinches chirm, boars girn, starlings murmur, geese creak. The seemingly slow, ambling walk of bears was referred to as "slothing."

Ordinary life in the past had an intimacy with other species that today we mainly associate with trained biologists and dedicated naturalists. Many plants and animals were burdened or

blessed with superstitions, such as claims that the houseleek*
would protect your home from fire or that toads drank milk from
cows' udders, yet there's no denying the daily communion that's
revealed in such forgotten facts as that homemade cowslip-flower
wine was once a popular beverage or that "the usual manner" of
making a salad in seventeenth-century England was "to take the
young buds and leaves of almost everything that groweth, as in
the gardens as in the fields." Into the nineteenth century it was
typical of men to be able to imitate the cries of game animals, and
many birds once had common names that suggest observation or
even neighbourliness: the species that today are called shrikes
were formerly better known as "butcher birds" for their habit of
impaling captured insects on thorns for storage, while today's
chaffinch and reed bunting were known as Jack Baker and Bessie
Blackers, respectively.† When Benjamin Franklin objected to the
bald eagle as a U.S. national emblem, it was because he knew the
bird from real life—its almost simpering cry and tendency to
steal or scavenge; it was, he said, "a bird of bad moral character."
The older names of flowers involved a similar degree of awareness; we might guess at the qualities of plants called hound's piss
and goodnight-at-noon, but it took real intimacy to name a flower
courtship-and-matrimony: its sweet scent fades after picking.

Musicians still write songs about the meadowlark, though few
people know the bird's song, often including, I suspect, the song-
writers. We use phrases like "bald as a coot" or "thin as a rail"

---

* Also known in the past as welcome-home-husband-though-never-so-drunk, though the
root of the name is unclear.

† These common names are taken from the U.K., where one species of shrike (the red-
backed shrike) has since been extirpated, and the reed bunting is listed as a priority for
conservation.

without knowing that a coot is a type of waterfowl with a smooth, white forehead, while a rail is a slender, secretive bird built to slip through thickets of marshland reeds. You might argue that we have forgotten such facts because they are no longer useful in the way they once were. Is that true? Is a meadowlark's song—which never had much to offer to our daily survival—less meaningful today than it was to our ancestors? Far more likely is that the meadowlark and the coot, the shrike and the bear, the weasel and the lion, are simply no longer a part of our lives.

◆ ◆ ◆

It has been said that a nature writer is a person who sees the same things as the rest of us but thinks he sees them better. There is a grain of caustic truth in that, but it hardly decides the argument. The issue today is not whether you see heaven in a wildflower, but whether you look at the flower at all.

To many, the idea of paying deliberate attention to nature may sound ridiculously old-fashioned. So is breathing, I suppose. An awareness of nature is not first and foremost a sentimental or spiritual practice, but a profoundly realistic one—a way of binding ourselves to the simple truth that human beings depend on ecological systems for our survival. Awareness is a countercurrent to the feedback loop of modern life. Pay attention, and we will value nature more. When we value nature more, we work harder to reverse its declines. Reverse the decline in variety and abundance, and nature becomes steadily more fascinating, more spectacular, more meaningful.

Awareness can be its own reward. One particularly endless February, when the grey and damp of the season had crept into life itself and good news seemed to have gone out of fashion, I noticed that the heads and necks of glaucous-winged gulls were changing, almost overnight, from the smudged brown of winter to the waiter's-apron white of breeding season. The traditional first sign of spring—the arrival of the first robin—was weeks away at most northern latitudes, but here was a more subtle, much earlier reminder that, yes, one day the sun would again beat down upon our backs. There is much to be gained and nothing to lose in these small acts of reconnection.

Nature remains a more hopeful place than the news about it might suggest. I recently joined three professional biologists for twenty-four straight hours of birdwatching (or birding, as aficionados call it, because the birds are often identified by sound rather than by sight). We started at 1 a.m., climbing high into the mountains in order to spend the day descending through every possible kind of habitat on our way back to the valley floor. By the following night, we had encountered 117 species of bird. The biologists were disappointed, but to me, it seemed miraculous. *One hundred and seventeen species.* It was more kinds of bird in a single day than I had knowingly seen in my entire life. They were everywhere, from spruce grouse pecking across the snowfields to an enormous great horned owl, outraged we had discovered him in his canyon lair.

There was a time when religious scholars sought to relate every species to the primacy of human beings—lice are our incentive to cleanliness, deer keep our meat fresh until we need it, horse shit smells sweeter than other turds because horses are chosen to live alongside us. For the most part we have left such thoughts behind, yet the way we shutter ourselves away from nature has much the same effect today, making it easy to believe that only our own species is at the centre of creation.

It's a difficult world view to sustain in the presence of the ruby-crowned kinglet, a bird that weighs less than a handful of coins and sings in forests so cold and high that no human culture in history has ever lingered there for long.

The biologists and I didn't only see birds. We saw bats and beavers and a pine marten. We saw snakes and two black bears. We saw what is not meant to be seen: the twin tips of a mule deer's ears where it hid in a stand of cattails, and a doe in secret stillness on her day bed. And we were able to see with our own eyes the vulnerability of so many creatures: the way that Lewis's woodpeckers appeared only in a solitary gully of wildfire-blackened trees, or cliff swallows gathered the damp clay to mud their gourd-shaped nests from a single puddle between a highway and a parking lot.

So much life, and such precariousness of life. In only a single day of careful observation, the wild landscape came to seem infinitely more alive, more abundant, more full of purpose than I had remembered—and because of that, more worthy of care. It remains a beautiful world, and it is its beauty, far more than its emptiness, that can inspire us to seek more nature in our lives and in our world.

◆ ◆ ◆

In a final kind of experiment, I spent a month taking daily swims in the ocean. It's odd that I had never done something similar before—for twenty years, I've never lived more than a half-hour's walk from salt water. Yet I sometimes go weeks without truly noticing the Pacific, which, after all, is the world's largest ocean, at its widest five times the diameter of the moon, at its deepest more than two kilometres deeper than Mount Everest is tall.

The beach where I swam every day was far from home; it happened to be the point where the Guadalquivir River spills into the Atlantic Ocean in southwestern Spain. The shore there was a perfect illustration of where the human relationship to the sea has taken us: Other than crowds of people sunning themselves, there was not a lot of life. After two weeks of afternoon swims, I had seen precisely zero fish of any size. I explained this away with the fact that the water was silty and the colour of caramel, but the birds, too, passed by mainly in the distance, and even seashells amounted to just a scattering on the sand. The main fish stocks that run between the Guadalquivir and the ocean— various species of eel, shad, mullet and lamprey—have shrunk in their distribution through the river and its tributaries by an average of 77 percent in just the last fifty years. The biggest of those fish, the European sea sturgeon, which can reach nearly four metres in length and outweigh me by five times, is at extremely high risk of extinction in the wild and has been extirpated from the Guadalquivir River. Typical cuisine in the area includes batter-dipped flatfish so small you eat them bones and all; shrimp so small you eat them without peeling off the carapace; and snails so small you eat them without pulling them out of the shell. I found a dusty book at the local library with photos of the fish market from the 1940s and '50s, and the fish were larger, but not by much. That's how far baselines have shifted in much of Europe and Asia: the golden years for fishing were a thousand years ago.

We retain the myth of the sea as boundless, so large and powerful that we are like specks at its edge—"Man marks the earth with ruin, his control stops at the sea," as Lord Byron wrote. Modern ecologists argue that this persistent belief is a delusion that blinds people to a worldwide crisis in the oceans. I had heard those scientists' words, but had never acknowledged their meaning until I made those daily visits to the shore. Looking down the beach at the tide of humanity in their bright swimsuits, I had a sad epiphany: the wild ocean has been tamed.

Then one afternoon I descended the cobbled streets to the now-familiar seaside and found it was no longer familiar. A full-moon tide had washed in and rearranged the beach. The whole bay, in fact, had taken on a different aspect, a new pattern of sandbars and pools. The sea had gently, beneath a clear sky and light breeze, moved the land around with the force of an aerial bombing run. On shore there was no hint of any change in the seasons—no twilight goosebumps, no lowering of the sun's gaze. Yet the sea was on a new path. It trembled with chop; *el poniente*, the west wind, steepened the faces of the waves. The tourists began to retreat. When a school holiday ended, the beach turned nearly empty overnight. The next day—the very next day— I waded into the water and was startled by a shoal of silver fish. I swam, but somewhere in the murky water I cut my hand on something hard and spiny, and crawled anxiously to shore, suddenly afraid of sharks. How did the fish know? How did they know to return at exactly this time, their schedule precisely tuned to our own? When we look closely at the living world, what we ultimately see is the mystery—we see how much we cannot see.

After that, the ocean came steadily back to life. More little fish appeared, and bigger fish to chase the small ones. Shells washed onto the beach. Tiny shorebirds arrived, running back and forth at the edge of the water as if afraid to get their feet wet, and then gulls—how strange to realize that even gulls had been a rare sight until then. A tall, white egret appeared, and then the rare Spanish imperial eagle. The beach bars folded their tents and at last I was the only swimmer left, though old men and women still dotted the shore, studying the sea as if it was the only thing on earth worth thinking about. And on the day before I left for a colder and more landward place, I took one final swim—just a way of saying farewell to a new friend. I was wading ashore, already feeling nostalgic, when some creature lashed out from the sand, stung my ankle, and then zigzagged off into the murk. Nature may not be what it was, no, but it isn't simply gone. It's *waiting*.

Chapter 6.

GHOST ACRES

———— ♦ ♦ ♦ ————

G reat Britain is not a wild place, but for all the black panthers. A colour variant of the familiar spotted leopard, black panthers turn up across the whole of the island, from the heathery moors of the Scottish Highlands to the chalk cliffs that face across the English Channel to France. By the most informed estimates, there may be one hundred or more sightings each month. For animals that can weigh as much as a heavyweight boxer and are feared man-eaters in other parts of the world, the British panthers are remarkably deferential, subsisting mainly on deer and rabbits rather than people or even livestock. Other big cats are seen in Britain as well, such as puma and lynx, but the vast majority of reports involve the panthers. According to one tracker, there are perhaps twenty black panthers in the southern county of Dorset alone, despite a human population density approaching 300 per square kilometre.

More than a dozen books have been written about Britain's big cats, which are darlings of the media—even beers have been named after them. Still, there is no conclusive proof that they actually exist. It's not impossible that they do: The panthers are said to be feral cats that have escaped from zoos or circuses or been released by owners who've grown tired of caring for large, demanding predators. At any given time, there really are many such animals in private hands; a 2006 survey of Britain's licensed exotic pets turned up 154 big cats, including fifty leopards. Still, the sceptic awaits a convincing photograph, a decisive video, a free-roaming panther found dead at the side of the road.

Great Britain is one of the least forested and most crowded regions in Europe—you can't hide a yeti there the way you might in the Himalaya mountains, or a sasquatch the way you can in the Canadian wilderness. But a black panther can fade into the smallest patch of shadow. They are famously silent, reclusive and hard to find. They hunt by night. And they are certainly fierce and frightening enough to epitomize the wild. "Either massive numbers of country people are experiencing social psychosis, or there is something out there that is worth investigating," says Alayne Street-Perrott, a geographer with Swansea University in Wales. It may be, though, that it is the landscape of the imagination that should be investigated. Whether or not black panthers are lurking in the British countryside, it is clear that a lot of people want very badly to believe that they are.

The natural world of the past is not simply gone and forgotten; in many ways it is still with us. The presence of absence is an idea that dates back at least to Plato, and is instantly understandable to anyone who has traced a family tree and seen the patterns of his or her own life reflected in the personalities, historical wounds and turning points of distant ancestors. To recognize that what has been lost is a part of what remains, however, still leaves questions of scale and character. How large an absence are we talking about? Where do we see its effects? What is the complete inventory of the missing? The answers to these questions not only shape the way we measure the world around us, but also help reveal the character of nature itself—including human nature.

Of all the regions on earth, Britain, which is geographically the island that houses England, Scotland and Wales, may tell us the most about how we accommodate the diminishment of the living world. The British countryside is deeply loved, globally influential—and almost entirely unnatural. In distant times, Britain was a peninsula joined to the rest of Europe, so that by the time it was ringed by rising seas, it was full of Eurasian animals, including such megafauna as lions, hyenas, woolly mammoths and woolly rhinoceroses, cave bears, bison and the enormous Irish elk.* Unable to migrate southward in periods of intense cold, some of these species may have been wiped out by ancient shifts in climate. As with so many of the world's lost megafauna, however, most of these species' disappearance from Britain is typically blamed on a combination of climate change and human influence. It's hard to overlook the strain of denial that runs through this claim—when we say that both climate and people drove the animals to extinction, we really mean that *we did it*. Before the arrival of people, the megafauna had weathered the coming and going of ice ages for millennia.

---

* The Irish poet Seamus Heaney described the skeleton of an Irish elk that he saw in a museum as "an astounding crate full of air."

Cause and effect become clearer with time. Five thousand years ago, Britain was covered in forests that are now pleasingly remembered as "wildwood." There were sprawling oak woods, stands of birch, pine and hazel, and forests of lime, which is now uncommon as a naturally occurring species even as individual trees. As much as 50 percent of those woodlands were already gone two thousand years ago. By 1900, Britain's original forest cover had dipped below 5 percent and has since recovered to around 10 percent—there's that number, again. Today's woodlands are heavily managed; nothing that can truly be called a wildwood remains on the island.

It was a similar story for the animal kingdom. Britain was once home to the aurochs, a species of wild cattle that probably spent most of its time in marshes and bogs. Do not picture the sedate, domestic modern cow: In Roman coliseums, bull aurochs proved powerful enough to kill tigers and bears. In Poland, where the last aurochs on earth, a female, died in the Jaktorów forest in 1627, the animal is said to have been fast enough to spin around and toss its own dung with its horns before it could hit the ground. Scholars at Corpus Christi College at the University of Cambridge still drink from an aurochs horn on special occasions, but aurochs were wiped out in Britain two thousand years before the species was declared globally extinct.* Other early losses include reindeer, as well as *Alces alces*, the animal known in Europe as the elk and in North America as the moose. Brown bears—the same species as North America's grizzlies, though typically smaller—were probably extirpated from England, Scotland and Wales during the days of the Roman Empire.

---

* The horn is thought to have been a 14th-century gift from Germany.

Wild boar and beavers followed in the thirteenth century, the Eurasian lynx in the 1500s and wolves by the early 1700s. By that point, Britain had lost all of its large wild animals but deer and red deer, the latter known to North Americans as elk. Deer watching is now a popular pastime.

The fall of the big beasts is far from the whole story. The environmental historian I. G. Simmons describes the years since the Middle Ages in the most British of terms: "There was a widespread lack of generosity towards wild birds and small mammals." Indeed. In the 1600s, the great auk was still abundant enough in some places that five tonnes of auk meat for salting could be taken in half an hour, but it had already been wiped out as a breeding species in the British Isles. The red kite, a scavenger and bird of prey once so common that children fed them bread and butter in the London streets, was extirpated by about the 1870s and was absent more than a century before it was reintroduced beginning in 1989. The story is similar for a long list of other species, among them the white-tailed eagle, bustard, osprey, capercaillie (a type of grouse), goshawk and wryneck woodpecker—each was eliminated from Britain, later to recover only slightly if at all through either assisted reintroduction or natural recolonization by birds from the continental mainland.

At the enthronement feast of the archbishop of York in 1466, the following birds were served: 400 swans, 2,000 geese, 3,000 mallards and teal, 204 cranes, 204 bitterns, 400 herons, 400 plover, 2,400 ruff, 400 woodcock, 100 curlew, 4,000 pigeons, 104 peacocks, 200 pheasants, 500 partridges, 1,200 quail and 1,000 egrets. The abundance of stilt-walking cranes, herons, bitterns and egrets on this menu suggests marshes and fens more full of life than any known on the island today. The Eurasian bittern was once familiar enough that the town of Bemersyde in Scotland takes its name from the males' booming call—like a tuneful foghorn—which was considered a harbinger of summer across northern Britain. The bittern is now seriously endangered on the island.

One commentator from that era defended the slaughter as "innocent wars"—fought against "birds and beasts alone." Britain is so well known for its role in the North American fur trade that it is often forgotten that the island once had its own. By the fourteenth century, fur-bearing mammals were already scarce enough that the right to wear fur was restricted to royals and nobles, and squirrel was a popular pelt—an aristocrat's bedspread could require 1,400 squirrel skins. Later, church wardens were deputized to pay bounties on a wide range of species that had been declared vermin, including foxes, polecats, stoats, weasels, otters, hedgehogs, rats, mice and moles. In a dozen years during the mid-nineteenth century, hunters on a single Scottish estate shot more than a thousand of the targeted animals, including 198 wildcats, or about half as many of these native felines—beautiful tabby animals with bottle-brush tails—as still exist on the whole of the island today.

Britain's environmental history can approach the tragicomic. At a "floral fête" in the spa town of Cheltenham in 1933, a prizewinner's bouquet contained twenty-two marsh helleborine orchids, then known to exist only in a single, three-acre bog. A funding drive in 1925 failed to protect a breeding ground of the Kentish plover from being converted into a golf course, and the bird was extirpated. More recently, Britons have worried that yet another species is disappearing from their shores. Black rats have called the island home for two thousand years, but today, the species that brought the Black Death to Europe has nearly vanished from Britain. Non-native brown rats—introduced from ships in the eighteenth century—have driven the non-native black rats—introduced from ships in the first century—to a last redoubt on the Shiant Isles off northern Scotland and to occasional appearances in British ports. "I'm just pleased to have seen a black rat," a local rat-catcher told the BBC after a rare encounter in Cornwall in 1999. The fur, he said, was "like velvet."

Much of this history is still recorded on the landscape. Place names, also known as toponyms, are one record of what used to be. A map can't always be trusted, of course: a place can be called Tiger Mountain because it's striped with white rock, or Wolf Creek because a man with the surname Wolf used to live there. Often, though, the connection is plain. Britain lost its beavers but is littered with places named after them, from Beverley (beaver glade) and Bevercotes (beaver homes) to Beverstone, Beversbrook and Beaver Hole. On the other side of the world, the state of California has more geographical features named for bears than for any other animal, including two hundred references to grizzly bears specifically. Grizzlies appear on the flag of "the Golden Bear State" and give their name to the University of California's varsity sports teams, yet the closest wild grizzlies to California can now be found eight hundred kilometres from the state's northeastern boundary.* There are bear rivers and bear mountains, bear meadows and bear canyons, bear gulches and bear buttes—a poem written over the landscape for those who know how to read it.

---

* California's flag may have another ecological reference. The bear is shown standing on strangely hummocky soil; some biologists believe these lumps represent the billions of pocket gopher mounds encountered by early pioneers. California's settlers waged war on the gophers as vermin, though they remain abundant in some areas.

Place names are a measure of the relationship between people and their surroundings. In 1947, a study of the Pamue people in Equatorial Guinea examined nearly 3,500 toponyms, many translating into such singsong terms as Pass Quietly Elephant. The Pamue proved to have a strikingly deep day-to-day awareness of biological diversity: Among other species large and small, their place names record forty kinds of mammal, including nine distinct primates; fifty-seven varieties of fish; and at least 178 species of wild plants. By comparison, one toponym researcher remarks that the place names of the entire English-speaking world might tally "a score or two" of native plants. By the time the British began to influence global culture, they had been distanced from wild nature for centuries, if not millennia.

The land itself remembers. In some cases, the hedgerows that mark the boundaries of many British fields were first established more than two thousand years ago, and old wild-wood plants and animals have managed to survive in them. Some, like hawthorn and holly, two common hedge plants, contain mysteries. At a glance, hawthorn appears to be well protected from browsing animals, given its long, stiff thorns. Curiously, though, the thorns are spaced widely enough that Britain's deer can easily eat the hawthorn's leaves. Holly, meanwhile, is a plant of two moods: its lower leaves are famously spiky, for protection, while its upper leaves are smooth-edged to create more surface area for converting sunlight into energy. What's strange in this case is that the shift from prickly to smooth takes place about five metres off the ground—far higher than necessary to guard against today's browsing animals. Hawthorn seems oddly under-defended, while holly is weirdly over-defended.

The solution to these riddles appears to lie in the deep past. Both plants evolved at a time when huge, plant-eating animals still roamed the world. Hawthorn is not adapted to defend against deer, but against megafauna. Animals like ground sloths, which resembled giant bears but were committed vegetarians, could not afford to daintily browse on individual leaves. Instead, they had foreclaws as long as a person's forearm, designed to hook thick clusters of vegetation toward their mouths; hawthorn's spines served to discourage such damaging browsers. Similarly, holly is designed to ward off animals tall enough to feed through the windows of a second-storey apartment, despite the fact that the plant hasn't encountered such threats for thousands of years. Holly and hawthorn are memory incarnate. They are ecological ghosts, manifestations of a world that no longer exists.

Ecological ghosts can be found almost everywhere on earth. Buy an avocado at the grocery store, and you're purchasing a seed designed to pass through a digestive tract larger than that of any living animal in the avocado's native habitat today. The same is true of the osage orange, a bizarre fruit that looks like a cross between a grapefruit and a human brain and piles up to rot at the base of osage orange trees across eastern North America. The osage orange is not eaten by modern man or beast, but fits the fossilized molars of mastodons the way soap sits in a soap dish. A survey in Brazil found 103 plants that showed adaptations to the past presence of megafauna. With no living animal able to effectively eat and disperse their seeds, some of these plants are now uncommon; others, such as the cacao bean from which chocolate is made, rely on cultivation by people. At least one, the tucum palm, is now found almost exclusively on riverbanks. Where once its seeds might have been carried far across the landscape by roving megafauna, it now relies mainly on a huge, fruit-eating fish called the pacu; the pacu, meanwhile, is threatened by a poorly regulated commercial fishery.

Perhaps the most extraordinary example of an ecological ghost is the pronghorn antelope, the fastest land animal in North America. At high speed, pronghorn can cover nearly ten metres between each footfall, meaning they can run an American football field in 3.5 seconds while touching the ground only ten times.* When breathing hard, they consume oxygen nearly as efficiently as a bat—pronghorn are, in a sense, built to fly without wings. Their major potential predator on the North American plains, meanwhile, is the coyote, which is laughably outmatched by prey that can accelerate from zero to one hundred kilometres per hour in a matter of seconds and sustain seventy kilometres per hour over long distances. Even a pronghorn fawn can outrun a coyote. The best current explanation for the overbuilt nature of the pronghorn is that the animal evolved among much faster, more explosive predators. Until the arrival of human beings and the fall of the megafauna, North America was home to lions, hyenas, supersized wolves and long-legged bears. Then there was the American cheetah. They were larger than the cheetahs that survive in Africa today, which are the fastest land animals on the planet and have been clocked at speeds of 102 kilometres per hour—just a little faster than a pronghorn.†

---

* This is made possible by a compact body that powers long, lightweight legs, a physiological design described by one pronghorn expert as "four chopsticks in a bratwurst."

† Asia also has cheetahs; they once ranged over a huge territory from the Mediterranean coast to India, but now survive only as a critically endangered species in Iran. In Africa, they were formerly numerous and widespread; they've lost 76 percent of their historical range on that continent.

Ecological ghosts are more than just interesting quirks; in some cases, they may be dramatically shaping the planet. One theory for the reason that much of the American West is choked with woody shrubs—much despised by ranchers because it makes poor fodder for cows—is that the big browsing animals that once beat it back, making way for grasslands, have been gone for thousands of years. If so, then not only individual species but whole landscapes can be thought of as relics of another time. Many of the ways we relate to the natural world, too, reflect changes that took place long ago. Looking again at Britain, the wholesale price of Atlantic cod—once considered "pauper's food"—can be twice as high as the cost of prime fillet beef. Cod, once so easy to catch that the first European fishermen in the Americas didn't even bait their hooks, have crashed nearly everywhere, and British fish and chips are now typically made with cod from Iceland or the Arctic Ocean—if they're made with cod at all. Britain's home cod fishery failed in 1920, and one historian has pointed out that more Britons today work in lawnmower manufacture than in the net fishing industry—this on an island nation, surrounded by the sea.

Like other European countries with devastated fisheries, Britain now buys the right to fish off other nation's coasts. These faraway waters are a part of Britain's "ghost acres," the distant reaches of land and sea that help make the island wealthy and populous, but without which it could not sustain itself. In the 1840s, only 5 percent of British food was imported from anywhere farther away than Ireland. Having passed through a frightening World War–era bottleneck when up to 70 percent of food was coming from overseas, Britain's governments have focused on improving the United Kingdom's self-sufficiency—but 40 percent of what Britons eat still comes from outside sources. Wood imports began in the year 1230, and by the time Scotland's capital city of Edinburgh underwent a major expansion in the early nineteenth century, all of the necessary timber was coming from the Baltic states and Scandinavia. In the 1740s, London was lit by five thousand whale-oil lamps, but even at that time, all of the whaling was done far from home ports. Faced in part with a domestic energy crisis, Britain became the world's first economy dependent on a non-renewable resource—coal—and then only reluctantly. Even feathers had to be imported; in the first ten years of the twentieth century, Britain shipped in six thousand tonnes of them, including, during a 1902 fad for feathers on women's hats, the equivalent of nearly 200,000 herons—one of the birds that had formerly been common in the island's marshlands.

Even the deep love of nature that many associate with the people of Britain, and which has so greatly influenced the rest of the world, is touched by the haunting. During the industrial revolution—which laid waste to many of the island's remaining forests, witnessed the invention of both urban smog and suburban sprawl, and polluted some rivers so badly that their water could be used in place of ink in fountain pens—a countercurrent emerged among British poets. I don't even need to use their first names: nature appreciation in the tradition of Blake, Wordsworth, Coleridge and the rest of the Romantics has called on us to see the face of God in every grain of sand ever since.

"Wheresoe'er the traveller turns his steps, / He sees the barren wilderness erased, / Or disappearing," wrote Wordsworth in 1814, though it is not for such warnings that the Romantics are remembered. Instead, we celebrate the poets for seeing the sublime in a wild landscape that others found dark and savage. It's been said that the Romantics taught us to see the wonder in a sunset, and while this is surely an overstatement, their love of field and stream helped inspire the global ideal of wilderness and laid a foundation for the modern appreciation of such seemingly charmless creatures as the hagfish, the warthog and the vampire bat. Yet the Romantics' love of the wild was ironic: It was a reaction against the destruction of nature, but also a product of that destruction. When the boy named William Blake walked out of London to the fields of Peckham Rye and first saw "heaven in a wild flower," a moment that could reasonably be considered the dawn of the romantic age, he lay down in a tamed landscape that could no longer endanger even a child. If the Romantics fomented a revolution in the way we perceive nature, it was in part because they had opened their eyes to a new reality: almost every threat posed by the wild landscape had been vanquished. By the early nineteenth century, Britain was much as we know it today—a deforested island, its fauna largely reduced to butterflies, birds and hedgehogs.

The pattern repeated itself on the American shore. Henry David Thoreau wrote *Walden* from a Massachusetts forest already emptied of large and fierce animals. Annie Dillard's *Pilgrim at Tinker Creek*, one of the most influential books of modern nature writing, plays out in a similarly denuded Virginia, and even Edward Abbey, that singular voice of wildest America, went to his deathbed never having seen a free-living grizzly bear. Such versions of nature still inspire wonder. In fact, one might argue that the works that have brought us closest to nature have *depended* on that wilderness being safe enough to approach and feel at peace in. But a greater truth should be foremost in mind: Nature is not a temple, but a ruin. A beautiful ruin, but a ruin all the same.

Today, there are efforts to rewild Britain, though the steps are necessarily small ones. Many Britons are fiercely attached to the wide open countryside they have always known, so that even the reintroduction of trees has, in places, been fiercely contested. One of the most ambitious reforestation projects is taking place in Glen Affric in the Scottish Highlands, an area of 1,500 square kilometres with no through-roads. I crossed the valley from end to end, spending a night in Great Britain's most remote hostel, and while the Caledonian pine is a magnificent tree,* spreading its gnarled limbs into the endless wind, no one who has passed time in deep wilderness would mistake Glen Affric for such a place. It is a beginning, a turning. Meanwhile, several beavers taken from Norway were released in 2009 in the Knapdale Forest on Scotland's west coast, an easy drive from Glasgow. The five-year trial reintroduction, closely monitored, came at the end of an official process nearly fifteen years long. Many legitimate concerns were raised, from the potential for beaver dams to block the spawning runs of fish to the fact that the beavers were returning to an island that lacks any natural predator large enough to control their population. The most common reason given for active opposition to the return of the beaver, however, was a lack of interest. Having lived four hundred years without beavers, many people saw no reason to undertake any effort, expense or risk on their behalf.

"People in this country are kept in ignorance of what their landscape represents," says Mark Fisher of the Wildland Research

---

* It also happens to be the historical "plant badge" of my own ancestors, the Clan MacKinnon.

Institute at the University of Leeds in northern England. Fisher is among those who hope to eventually see the reintroduction of species as demanding on the human psyche as the grey wolf and Eurasian lynx, though he acknowledges that the rewilding of Britain must first begin with trees, with beavers. He's acutely aware of the history of nature in his country and its various social and ecological costs, and yet when I ask him which of the losses he feels is Britain's greatest warning against allowing the whole of the planet to drift down the same path, his answer surprises me. "I think you lose a lot of human freedoms," he says. "The human freedom to experience wild nature."

The lone person on a wild landscape is a baseline of human liberty, a condition in which we are restrained only by physical limits and the bounds of our own consciousness. It is for this reason, perhaps, that so many of us are drawn to nature as a counterpoint to the world of regulations and traditions, grids and networks, that we live in day to day. Yet when that nature is a landscape of straightened rivers, cleared forests and drained wetlands, beneath skies emptied of birds, it is only more of the same, its meanings already decided by other people in other times. It was surely this truth that inspired the American philosopher Aldo Leopold to write his most enigmatic line. In the final words of his 1949 eulogy to the Colorado River, which would eventually be dammed and diverted so completely that it no longer flowed into the sea, Leopold writes, "Of what avail are forty freedoms without a blank spot on the map?"

Fisher still tries to find that freedom. He follows deer trails, crawls through undergrowth, climbs to mountain ledges that even grazing livestock cannot reach. Often, he does so by trespassing; despite new laws that allow people in England to roam freely on some undeveloped private lands, the public still has access to less than 10 percent of the landscape. His greatest challenge, though, is simply that there are not many places in Great Britain that are not unmistakably marked by human hands.

When he can, Fisher travels to North America, where he seeks out landscapes that serve as reminders of what Britain might have been. Once, in Yellowstone National Park, he saw a pack of wolves emerge from a forest's edge into an open meadow—he fell down on his knees, he says, from the impact of that vision. But it was the wildwood, the simple trees, that broke him. He was hiking in New Hampshire's White Mountain National Forest when he suddenly stepped out onto a rocky overlook. From there, he had a view across more than three thousand square kilometres of woodland. "I just cried my eyes out," he says. "I mean, you can't have those experiences in the U.K., because they don't exist."

Chapter 7.

UNCERTAIN NATURE

———— • ♦ ♦ ♦ • ————

In around 1620, a Flemish artist named Frans Snyders completed a painting called *Fish Market* for the wall of a tax collector's house in the Netherlands. The canvas is unlikely to make you want to swim in the waters off Antwerp. Dead things of the sea lie heaped upon a table, and there are pincers and tentacles and many, many eyes. To one side stands a fishmonger—a portrait of a man among monsters.

Look awhile longer, and you begin to see that all those terrible creatures are only ordinary animals: cod, herring, halibut, a bottlenose dolphin, lobsters. It's just that nearly everything is oversized. There probably never was any actual fish stall in the Netherlands with such a madcap sprawl of seafood; at the same time, the image is far from fantastical. Every species but one that Snyders depicted was once fished in the Wadden Sea off the Netherlands coast. The lone exception is a red-footed tortoise, a native of South America that must have made it to Europe on a Dutch trader's ship.

It was thought at one time that artists had better vision than the rest of us—that through painstaking hours attempting to capture the outrage in a squid's eye,[*] a painter like Snyders could develop magnifying vision. Science ultimately showed that it was impossible to improve eyesight through repetitive exercises, though George Perkins Marsh, for one, couldn't bring himself to believe it. Struggling with bouts of near-blindness as a boy, he was unable to read books and instead studied the landscape around him. How well your eyes worked, he discovered, mattered less than knowing what you were looking at. "Sight is a faculty," Marsh wrote. "Seeing is an art."

It is the same with the natural world today. We cannot hope to look at a wild landscape and see its history with our eyes; only the mind's eye can capture it. The living world has been depleted, but it has also been *transformed*. This is the reality that Heike Lotze, a marine ecologist at Dalhousie University in Nova Scotia, sees in Snyders's painting.

Lotze grew up on the Wadden Sea and is one of those people who seems to carry an ocean breeze in her pocket wherever she goes. Her childhood home near Norden, Germany, was just behind a low berm that centuries earlier had been the first dike in the area to claim land from the sea, and is now a roadway with a name that translates roughly as Carrot Street. The high tide has not kissed the Carrot Street dike for a long, long time.

---

[*] Vegetarianism was in fashion in 1620s Netherlands, and one Snyders scholar has argued that this may have influenced the artist's sympathetic portrayal of fish, which he often shows twisted in death throes or staring with accusing eyes.

By the time Lotze was born, the coast was five kilometres away from her family's home, and the sea was held back by an eight-metre-high wall.

Thanks to Lotze, the historical ecology of the Wadden Sea is now one of the most exhaustively studied on the planet. The sea is a long, narrow strand of water that separates mainland Netherlands, Germany and Denmark from a chain of barrier islands known as the Frisians. The Wadden is the world's largest intertidal area, meaning there is no other place on earth where so much water gives way to land, and vice versa, between the tides. At high tide it is a shallow sea, and at low tide mainly a mud flat.

The Wadden formed about 7,500 years ago, at the end of the last ice age, and the first human settlers arrived not long afterward. It was probably a welcoming, if muddy, home. Despite receiving silt from several of Europe's major rivers, including Germany's longest, the Rhine, the Wadden Sea is thought to have had clear water, filtered in part by the stillness of its coastal wetlands and in part by a constellation of filter-feeding life forms. It was a strangely soft place—the Wadden has almost no natural rock besides erratic boulders set down here and there by retreating glaciers. The only hard surfaces on the ocean floor were produced by living things: huge beds of beach oysters and blue mussels, along with the strange, soda-straw reefs of *Sabellaria spinulosa*, a sea worm. Above all else, the Wadden was a true sea: a nearly complete natural system in and of itself.

Historical ecologists have a maxim for what typically happens when human beings move onto a new landscape: we eat the big ones first. The earliest Neolithic hunters and gatherers fed themselves mainly from the land, hunting boar and deer, beaver and otter, and looking to the sea mainly for ducks, seals, porpoises and stranded whales.* By the time of the Roman Empire, nearly every living land animal was on the table. Voles—rodents similar to mice—were an important food, but so was such unusual fare as foxes, wild cats, hawks and crows. Yet it wasn't until the Middle Ages, about 1,500 years ago, that the people of the Wadden shore, struggling to feed themselves from an increasingly empty landscape, finally focused on fish. Even then, they drew mainly from freshwater lakes and rivers. It took another seven centuries before the stocks of salmon, sturgeon and pike were in decline. Once again, human adaptability provided a way forward, with a hard turn toward the sea.

It was an extraordinary moment—a kind of rebirth. The wild abundance of the land was gone and forgotten. Yet in the sea there was wealth beyond imagining, and that treasury would help fuel the European empires that would go on to reach around the globe. It was a time of such strange sights as sailing ships designed not only to keep sea water out, but also to store it inside the hull, so that fish could be delivered, alive, to distant ports. It was a time when a baker might rise before dawn, light lamps fuelled with the rendered fat of whales and seals, and start cracking eggs taken from seabirds—thirty thousand eggs a year were gathered on the Wadden Sea's barrier islands for the Amsterdam bakeries alone. The markets were full of wild waterfowl, and the Netherlands' first cookbook

---

* One theory for why some whales strand themselves is to avoid drowning. They are air-breathing mammals, and when sick, wounded or near death from old age, may beach themselves rather than die inhaling water.

featured a recipe for porpoise pepper steak. The arrival of the harbour porpoise was considered as sure a sign of summer weather as flights of swallows were the heralds of spring.*

The first marine species to disappear from the Wadden Sea were the whales. By the early 1700s, the entire Atlantic Ocean population of grey whales had been wiped out, and the northern right whale could no longer be found anywhere in Europe. A pattern was quickly established: in the beginning, total plenitude, and within a few centuries, near eradication. Whales, dolphins, porpoises, seals, seabirds, haddock, houting, cod, rays, plaice, sole, dab, herring, anchovies, sprat, salmon, flounder, eel, lobster, even peat moss, even seagrass: nearly 90 percent of the Wadden Sea's major species are now classified as depleted. More than a fifth have been extirpated entirely.

Diking has shrunk the Wadden Sea to half its original size, to the point that the Rhine River is no longer said to flow into it. The delicate blurring of land and water is gone. Until one thousand years ago, settlers on the Wadden lived on mounds raised above the marshes or built temporary villages among shifting dunes. Today, the Wadden is a hardened coast, with more than seven hundred kilometres of human-built stone or concrete shoreline. The oyster and seaworm reefs are gone, and the eelgrass meadows have largely been replaced by an invasive species, cordgrass, that was planted to help stabilize the shore against erosion. The Wadden is so changed that its

---

* Porpoise was once a much-loved meat, to the extent that a writer in the *Transcripts of the American Fisheries Society* in 1885, having eaten broiled meat from a porpoise caught off Cape Hatteras, wondered why it had been forgotten. "The golden age of gastronomy was long ago, and in that time kings and other great persons looked upon porpoise as a delicacy of delicacies," writes the correspondent.

waters have not been clear in living memory; so changed that the remaining porpoises make their appearance not in summer but in winter; so changed that not one of the species that appears in Snyders's *Fish Market*—not a single one—is still commercially fished on the same coast today. So little remains of a natural system that was once nearly self-contained that at least one biologist has suggested the Wadden Sea should no longer be considered a sea at all.

For Lotze, the older Wadden Sea is a place she knows only from history and statistics, like a memory at the tipping point between remembrance and forgetting. Today, much of the Wadden is protected in national parks. With its fishing days all but done, the economic linchpin of the Wadden is tourism: people come to witness what they see as a wild ocean. Even the remaining fishermen, who bring in fish barely longer than their hands, often refuse to believe the history that Lotze tells them. "People think I'm just making things up," she says. "They are determined that there were never bigger fish in the Wadden Sea."

Recently, residents have become concerned that something strange is happening in their local waters. Fully protected from hunting, the Wadden's harbour seal population has rapidly boomed from scarcity to twenty thousand animals or more. In the past, the sea was home to twice that many seals, but to people today, it's like nothing they've ever seen. They describe the return of the seals as "unnatural."

◆ ◆ ◆

There are places where the human impact is overwhelming; no one expects that all eighty species of fish that once lived on or around the island of Manhattan, for example, might somehow still be swimming through the sewers of New York. But how deep is the overall scale of change, and how wide?

To begin to get a sense of this, imagine African elephants moving across a savannah dotted with trees. The scene is wondrous, the herd seeming to move more to the rhythm of clouds than to the hurried pace of human beings. The poet John Donne called elephants "the only harmless great thing," but then, Donne was not a botanist. A clan of two hundred African elephants will eat or otherwise destroy more than sixty tonnes of vegetation per day—an equivalent weight of hay would feed a herd of more than five thousand cattle. Of course, elephants aren't eating hay. They're eating almost anything green they can find, and if that means tearing down a fifteen-metre-tall tree to reach its highest leaves, then that is what they will do.

In the late 1800s, elephants were almost totally wiped out by ivory hunters in the Serengeti region of Central Africa. Today, the Serengeti is familiar from nature documentaries and the pages of *National Geographic* as a great, grassy plain—but that's not what the region used to look like. In the 1930s, much of it was a dense forest. Colonial administrators set the area aside as parkland in large part because it was sparsely inhabited—Maasai herders had abandoned the region because the bushy landscape harboured the tsetse fly, which can carry potentially lethal sleeping sickness. Then, in 1955, elephants began to return. Almost immediately, the forest started to disappear.

As the elephant population recovered in various locations across Africa, the same pattern repeated itself: woodlands went into retreat, and grasslands began to advance. For years, this was referred to as "the elephant problem" and considered unnatural—the outcome of too many elephants. Today, it is more often looked to as an example of how different a complete ecological system can be from one that is missing key species.

Elephants are known as ecosystem engineers, one of a list of animals—from beavers and sea turtles to earthworms and humans—that radically reshape the landscapes they live on. The large-scale changes these species create trickle down. Forests mauled by elephants are home to larger numbers and more varieties of lizards, for example; the Parker's dwarf gecko prefers these woodlands so strongly that they abandoned their home trees when scientists experimentally repaired the elephant damage. Most herd animals and antelope also prefer elephant savannahs, which provide few places for predators to hide and where the broken trees sprout new shoots close to the ground. Even elephant droppings are useful, offering moist hideaways for frogs in the dry season and, in one unenviable study of dung piles in Cameroon, spreading the seeds of ninety-one species of plant. Elephants also produce waterholes—I once watched a fully grown female work her way down into the mud of a wallow until only the tip of her trunk stuck out like a snorkel—and in at least one case are even known to make caves. On Mount Elgon in Uganda, elephants have been scraping away at salt deposits for perhaps twelve thousand years, resulting in caverns that reach as much as 120 metres into the hillside and are used as shelter by animals ranging from leopards to bats. They do most of their salt mining in the dead of night.

Remove just this single species, elephants, and you end up with a different environment. Now consider the fact that Africa was once home to ten million elephants, or twenty times as many as live there today. Note that, at the end of the last ice age, elephant-like animals roamed every continent except Antarctica and Australia. There were even dwarf pachyderms on many islands, from the Channel Islands of California to Wrangel Island in Arctic Russia.* Elephants, mammoths, mastodons and the like have disappeared from 90 percent of their Pleistocene range, and they probably affected their habitats in much the same way as modern elephants do in Africa and Asia. One mystery that ecologists struggle with, for example, is why there are grasslands in many parts of the world that have adequate soil and rainfall to support forests. A possible explanation is that elephants and other plant-eating megafauna kept the trees from encroaching and allowed prairie ecosystems to take hold. Even the far north, today a world of wet tundra, was once widely covered with a dry, grassy steppe—which may have depended for its existence on heavy grazing by mammoths. Supposing ancient mega-herbivores shaped only the world's grasslands (there's little reason to imagine their influence stopped there), we are already talking about nearly half of the earth's terrestrial surface.

One other species may have benefited in extraordinary ways from elephant engineering, and that is us. Among the most durable findings in the field of environmental psychology is that people prefer natural settings over the built environment. Among natural landscapes, however, we show the greatest preference for open spaces dotted with trees, with a little water nearby—picture the views from the high-rises that famously border Central Park in Manhattan; as the biologist

---

* Small woolly mammoths still existed on Wrangel Island when the Egyptians were building their first pyramids, disappearing only about 4,000 years ago.

E. O. Wilson puts it, "to see most clearly the manifestations of human instinct, it is useful to start with the rich." This predisposition has proven true in experiments across cultures and generations with a consistency that has given rise to the "savannah hypothesis," which suggests that we value such places because they resemble our ancestral home—the plains of Africa where human beings evolved.

Those plains are also precisely the type of landscape that elephants create. Indeed, we now know that the Maasai herders who had largely abandoned the Serengeti by the time it was declared a protected area had in fact lived there when it had elephants and returned with their cattle as the elephant population recovered. The most pronounced marks that elephants leave behind are their trails, tamped hard underfoot and perfected over centuries to find the line of least resistance through hills and valleys. Within memory, an elephant route that stretched across much of northern Uganda was considered the best road in that country, and, as with the buffalo trails in North America, some such ancient pathways have gone on to become modern highways or railroads. Archaeologists continue to argue about how it is that early humans spread with such remarkable speed around the globe once they finally left Africa. It's not hard to imagine that, often enough, we followed in the footsteps of elephants and other animals. The first human beings to arrive in these new worlds thousands of years ago were perhaps similar in at least one way to the European explorers who came in the age of sail: they discovered a world that had already been engineered by its inhabitants, and would forever be changed by the new arrivals.

◆ ◆ ◆

There is probably no species that doesn't leave behind a shadow as it burns out of existence. Even small animals can be ecosystem engineers. Indian crested porcupine burrows shape Middle Eastern deserts, for example, and even appear to affect relations between people and lions, tigers and leopards, which if injured in an encounter with a porcupine are more likely to start hunting humans; the infamous Leopard of Gummalapur, which killed forty-two people in India, was found to have two quills stuck in one paw. Honey bees, which are not native to the Americas or Australia, are endlessly promoting plants with flowers that they can pollinate at the expense of less accommodating species. Such ripple effects have come to be known as "cascades," and some of the most potent involve those flesh-eating animals we call predators.

In the 1990s, a riddle began to play out off the southwest coast of Alaska. Sea otters, which had rebounded from the fur trade so successfully that in places they may have equalled their historical abundance, suddenly began to disappear. Over the next ten years, the otter population plummeted by 90 percent.

To explain the vanishing otters, researchers eventually found themselves looking at a completely different era and completely different species. Whaling in the North Pacific from the mid-1800s onward gradually reduced stocks of the great whales—humpback, blue, right, grey and the wondrous bowhead, which can live two hundred years or more. After World War II, industrial whaling hit even harder, drawing down the still-abundant fin, sei and sperm whales. On average worldwide, whalers killed the equivalent of one hundred whales every day of the last half of the twentieth century.

In the 1970s, after the U.S. had banned all whaling in its waters, harbour seals began a steep and inexplicable decline. Then fur seals. Through the 1980s, sea lion rookeries collapsed. At last, in the 1990s, the sea otters began their own downward spiral. The culprit, several scientists finally proposed, was the killer whale. The name is not as apt as it sounds: not all killer whales kill whales, but some specialize in doing so. Whaling left them an ocean almost empty of whales, and, in the same way that people do, they soon began eating their way down the food chain. By the time they focused on otters, they were dealing with snack foods—it would take more than 1,800 otters a year to keep a single killer whale properly fed. At that rate, just four dedicated killer whales could have crashed the otter population.

The otters were not the end of the cascade. As they disappeared, another chain reaction started. Otters are predators too, and among their favourite foods are sea urchins, those bottom-dwelling creatures, usually in shades of green, purple and red, that look like balls of knitting needles. Urchins, in turn, are plant-eaters with a special fondness for the long, banner-like seaweeds that are the "trees" in the underwater kelp forests along North Pacific shores. Remove otters from the food chain, and the booming population of urchins soon grazes the kelp down to stubble. Even that is not the end of it. Change a kelp forest into an urchin barrens and you alter everything from the diets of bald eagles to the growth rates of barnacles to the height of the waves that strike the shore.

As with most mysteries, the case of the missing otters seemed simple enough once it had been solved. In reality, it took more than a decade to put the puzzle pieces together, and the solution remains contentious and hypothetical—one zoologist describes the science of ecological cascades as "the discovery of pattern and order against a backdrop of noise." It is the realm of the counter-intuitive. For example, common sense might tell us that if you eliminate large predators, then the animals they usually eat will increase. Yet in the West African nation of Ghana, the opposite took place when the killing of lions, leopards, hyenas and wild dogs unexpectedly caused olive baboons to switch from being mainly gentle vegetarians to being intelligent, organized and vora-cious meat-eaters, devastating prey animals. Who would imagine that the introduction of deer to a landscape might also end up wiping out black bears? On Anticosti Island, Quebec, exactly that occurred: the deer population exploded and browsed the island's berry bushes so severely that the bears starved to death.

How much of the world around us has been transformed to these extremes? I put the question to James Estes, a University of Santa Cruz ecologist and pioneer of cascade theory. It was Estes who, with marine biologist John Palmisano, first sorted out how a kelp forest could turn into an urchin barrens. But even the kelp forests themselves, Estes said, might have been the end results of even earlier human impacts. Centuries before the sea otter became the gold standard of the maritime fur trade, a huge but almost totally forgotten animal, the Steller's sea cow, was moving through the kelp in herds. Sea cows were related to the dugongs and manatees that today are found only in tropical waters—placid and bewhiskered, they resembled walruses with-out the tusks. Sea cows had once been widespread on both the Asian and North American coasts of the North Pacific, but dis-appeared from most of their range in prehistory, probably at the hands of human hunters. By 1741, when the species was first recorded in scientific literature, Steller's sea cows could be found only around the isolated Commander Islands of the Russian Far East. Twenty-seven years later, they'd been hunted to extinction. What a kelp forest might look like with hungry sea cows browsing through—the way herds of African elephants browse through savannah woodlands today—can only be imagined.

Big herbivores like the sea cow and elephant, and big predators like lions and sharks, rank among species that have endured the most severe declines since the emergence of modern human beings. Based on what he has learned about contemporary cascades, Estes estimates that human actions over time have altered the natural evolution of ecosystems on more than 90 percent of the globe. The cascading impacts of those losses over the millennia were probably gradual and essentially invisible, and today are mostly unknown and possibly unknowable. When Estes travels in wild places today, such as Yellowstone National Park, which is typically seen as a fragment of near-original American wilderness, he says that he no longer thinks to himself, *This must be the way things used to be*. Instead, he finds himself thinking, *I'll bet this is a hell of a lot different from the way it was*.

"So we're really talking about a different world," I say to him.

"A very different world," Estes replies. "A *very* different world."

◆ ◆ ◆

There is at least one more horizon of transformation to explore—a more personal horizon, if it's possible to say such things about non-human life.

We will need to revisit the elephants, this time in apartheid-era South Africa. In the early 1980s, the elephant population was swelling in Kruger National Park, and wildlife managers decided to dart numbers of adult elephants from the air and then shoot them to death on the ground. The killing often took place in plain view of juvenile elephants, which were then rounded up and sent to other parks and preserves, with about forty ending up a few hundred kilometres to the southwest in Pilanesberg National Park. It must have seemed like a logical if gruesome act of conservation: reduce overpopulation in one place and spread the wealth of the species to others.

More than a decade later, field biologists in Pilanesberg noted what they termed a "novel situation" emerging. Elephants were, for the first time in history, killing white rhinoceroses, which had been bred back from the brink of extinction. Between 1992 and 1998, elephants were suspected in the deaths of forty-nine rhinos—a massacre. The culprits turned out to be the orphaned young males from Kruger. The obvious conclusion to leap to would be that the elephants' berserk behaviour was rooted in the trauma they'd endured as calves. Ultimately, however, the investigation turned to a question of the animals' culture.

As they approach maturity, male elephants enter a rutting condition known as musth, during which testosterone floods their systems so fiercely that even their posture is changed. The adolescent males in Pilanesberg were entering musth too young and staying in it too long; one suspected rhino killer was finally culled after remaining in musth for as many as five months, a length of time that would be unusual even for a male twice its age. Under more natural circumstances—that is, in an elephant herd not comprised of transplanted orphans—the adolescent musth periods are cut short by apparently withering encounters with larger, older males. After standing down to a dominant bull, the rush of hormones in the younger male stops, in some cases in a matter of minutes. As a test, six older male elephants were introduced to

Pilanesberg, and the killing of rhinoceroses stopped. The out-break of elephantine violence was blamed on a lack of elders.

Africa's "behemoths"—the name given to the continent's huge, old elephants—were mostly wiped out decades ago. Elephant tusks grow throughout their lifetimes, and the largest tusks ever recorded, with a combined weight of two hundred kilograms of ivory, were cut from a bull that was shot on the slopes of Mount Kilimanjaro in the 1890s. In fact, evidence that human hunters target older animals as a general rule—they tend to be less alert and less dangerous than animals in their prime—stretches back at least to the Middle Stone Age. The trend continues: harvesting by human beings is the leading cause of adult mortality for a long and growing list of species.

We've brought an end to old age as a normal part of life in the natural world, and the effect could be so great that Anne Innis Dagg, one of the few zoologists to have written on the subject of elderly animals, argues that it may no longer be possible to know whether or not we are looking at "natural behaviour" when we observe the social order of animals. Elephants are one of the few species in which the importance of aging is slowly being acknowledged. During a 1993 drought in Tanzania, the elephant clans led by the oldest females suffered far fewer deaths than those with younger matriarchs—the herds needed leaders old enough to remember the distant waterholes that their own elders had led them to during droughts in the past. To maintain a mental map of those life-saving pools requires the continuous presence through the centuries not only of adult elephants, but elderly ones—severe drought strikes Tanzania only every five decades or so, and elephants' maximum lifespan is about sixty-five years.

Such impacts may not be limited only to big-brained, long-lived species. In the early 1990s, the fisheries scientist George Rose headed for the Grand Banks of Newfoundland to attempt to record, for the first time, the migration routes of cod in the northwest Atlantic. The knowledge was suddenly urgent: the fishery, long a symbol of the baroque abundance of which nature is capable, appeared to be failing. Rose predicted the likely migratory path based on water temperatures and sea bottom topography, and echo sounders proved his forecast was largely correct. The fish were there. But he also made an observation he had not expected.

As with the elephants, the last "behemoth" cod disappeared long ago; an Atlantic cod over ninety kilograms hasn't been caught since the 1890s. Still, it was the largest, oldest fish that seized Rose's attention on the echo sounder readouts. He could see them distinctly, individual black smudges at the head of every school. "Scouts," his team called them. In most of the schools, there were only a few such leaders left. Rose came ashore with many questions. What signposts did these scout cod follow through the vast, undifferentiated space of the sea? How did they determine where and when the schools would spawn? Was he really watching fish that had the wisdom and memory of years, that were keepers of knowledge passed down through generations?

Answers would not be forthcoming. In 1992, after centuries of overfishing, the cod stocks collapsed. You could still find cod on the Grand Banks, says Rose, but they were little things, few of them more than five years old. The elders were gone. And for the first time in five hundred years of written history, the ancient cod migration failed to take place.

Chapter 8.

WHAT NATURE LOOKS LIKE

———•♦♦•———

O ne of the most memorable scenes in the novel *Moby-Dick* takes place along what sailors once called the Line—the Equator—in the Pacific Ocean. There, late at night among unnamed islands, the crew of the *Pequod* kills a sperm whale and lashes it to the ship's side to wait until morning to cut up the carcass. In most places, this is a reasonable practice. Scavenging sharks will come—so many, in fact, that the sea will seem as "one huge cheese, and those sharks the maggots in it," as Herman Melville writes—but these can be driven away with whaling spades, which look like sharpened, flattened shovels on the ends of long wooden poles. On the faraway islands of the Line, however, "incalculable hosts" of sharks appear, beyond anything the whalers have ever known, their frenzy so wild that a shark disembowelled with a whaling spade might eat its own entrails, only to have them pass back out the hole that they spilled from in the first place. Even the dignified chief harpooner of the *Pequod*—the character Queequeg, who shaves with his harpoon, is covered in "unearthly tatooings," and is descended from the cannibalistic royalty of the imaginary island of Kokovoko—is rattled by the maelstrom. Lowered by lantern light to hack at the monsters from the ship's cutting platform, he finally climbs back on deck only to nearly lose a hand to a shark that has been hoisted aboard as dead. "Queequeg no care what god made him shark, wedder Fejee god or Nantucket god," he says, "but de god wat made shark must be one dam Ingin."

To those who know today's oceans, the scene seems to say more about Melville's writerly imagination than the true life of a sailor. Sharks are an uncommon sight on most of the world's coral reefs, which are scattered like pearls from the Red Sea in the Mideast, through the Indian and South Pacific Oceans, and onward to the coasts of Central America, the Gulf of Mexico and the Caribbean. Taken together, the oceans' reefs cover an area only as large as Italy, or about half the size of Manitoba. Three quarters of them can be found within fifty kilometres of human settlements, where sharks infesting the water like maggots would be sure to attract some attention.

For most of history since Melville's day, the world largely forgot about the distant reefs visited by nineteenth-century whaling crews. Ten years ago, marine biologists in search of the world's most pristine reefs began exploring what are today known as the Line Islands, a raft of remote South Pacific atolls mainly under the flag of Kiribati, along with three that are U.S. possessions. The most isolated of these islands, Kingman, is as near to being an ocean wilderness as there is today. To begin with, the atoll's thin strips of treeless beach, barely above the tide line, are uninhabitable, while its reef-protected harbour has never seen any heavier use than as a way station for Pan American World Airway's "flying boats" in the 1930s. Today, Kingman and its waters are a U.S. national wildlife refuge.

Enric Sala, now an explorer-in-residence with the National Geographic Society, was on one of the expeditions that revisited the Line Islands. The scientists knew the most remote atolls were different even as they approached them by boat, he says, because the air was full of the sight and sound of seabirds. "You go to inhabited islands in the tropics and the birds are gone, because people ate the birds or the eggs, or rats ate the chicks," says Sala. The team dove at more and more remote reefs, and were excited to be seeing sharks on every dive, so the first person to roll off the boat when they reached Kingman was the expedition photographer, hoping to shoot images of the reef in its most undisturbed condition. Instead, he quickly hauled himself back out of the water. "We can't dive here!" he shouted. "There are too many sharks, and they're too curious!"

In the end, they did dive at Kingman Reef. Everyone jumped in together, scattering the sharks, then waited for them to return, check out the divers, and finally move on to their ordinary routines. Only then could the scientists begin their work of surveying the reef's species. This time, the photographer would return to the boat with images that in some cases captured thirty sharks in a single frame. They had found sharks as they were known from the South Pacific of history, from *Moby-Dick* and seamen's logbooks recalling "sharks innumerable, and so voracious that they bit our oars and rudder," lagoons "infested with sharks," where one sailor was bitten on the heel in three feet of water and barely made it to shore as "a shoal of ravenous monsters" rushed toward the scent of his blood. One of the Kingman dive team later described his experience at the reef as a transfiguration into "a born-again biologist"—his eyes finally opened to what a natural reef really looks like.

"I don't think I would swim at a pristine coral reef," says Sala. "Underwater with our black wetsuits, now, that's different. But if you're swimming, like chop, chop, chop, and they see all your white belly—I wouldn't do that. It's a different world. A landscape of fear."

♦ ♦ ♦

The last, best hope to understand nature without the human imprint is in the sea. The hazards and technological difficulties involved in reaching the points most distant from land or farthest below the surface made the open ocean the last place on earth to be affected by our actions. We have no written record of the continents ahead of human settlement, but history is filled with accounts of briny waters that had rarely if ever before seen men. It's remarkable, then, that these reports describe a world beyond our current understanding.

In 2002, fisheries scientists at the University of British Columbia in Vancouver attempted to develop a computer model of past states of nature in Hecate Strait, a body of water off Canada's west coast that is famous for winter swells that roll in at heights of ten to twenty metres even without a storm blowing. The researchers divided the strait's known sea life, from whales to plankton, into fifty-one categories, and followed the data for each one back through time. In more than 80 percent of cases, those sea creatures were more abundant in

1750—earlier than any confirmed contact between the local native population and European or Asian explorers—than they are today. In fact, more than 40 percent of the categories were at least twice as plentiful in the past.

Put simply, the Hecate Strait of 1750 had more of *everything*. Not just larger numbers of the big, eye-catching species (two times as many whales and salmon as today, four times as many lingcod, sixteen times as many sea otters, and so on), but also huge populations of smaller creatures: more so-called "forage fish" such as herring and smelt, more shellfish, more shrimp, more crabs, more corals, more sponges, more seaweed. The trouble being that it's not supposed to work that way. "It doesn't fit with conventional ecosystem models, because if there were more predators, there should have been *less* forage fish. More cats, less mice," says Tony Pitcher, a pioneer of marine historical ecology who worked on the Hecate Strait model.

Similarly challenging findings have turned up almost any-where that people are looking into the question, from the South China Sea to the Mediterranean, from the North Sea to the Gulf of Maine. One possible explanation is that the assembled data overstates the past abundance of the oceans. Historical ecologists have been criticized for their reliance on the unscientific observations of early explorers, settlers and even naturalists, not to mention books like *Moby-Dick*—a work of fiction, after all. Yet to ignore these sources of information—which often include the only eyewitness accounts of any given place in the past—can impoverish our knowledge to a surprising degree. A groundbreaking 2006 effort led by the San Diego–based Scripps Institution of Oceanography to identify past versus present nesting beaches of Caribbean green sea turtles found references in 163 historical sources, ranging from Charles de Rochefort's 1666 *The History of the Caribby-Islands* to one of the earliest American novels, William Williams's semi-autobiographical *Mr. Penrose: The Journal of Penrose, Seaman*. Supported by modern nesting data and historical turtle-hunting reports, the study's authors cast new light on the species' supposed "recovery" from near-extinction in the early twentieth century; today's green turtle population, they estimate, is 0.33 percent of its former plenitude. Admitting that historical records are not precise, the researchers acknowledge that the true number could be higher—but then again, it could also be lower.

The Hecate Strait computer model was built using the widest possible range of source materials, including modern fisheries data, archaeological evidence, historical records, and local knowledge from both indigenous people and later settlers—and it still produced a confounding result. Scientists have an increasingly clear idea of what the oceans *contained* in the past, says Pitcher, but struggle to say what it might have *looked like*. He likens a computer model of nature to the abstract portraits painted by Pablo Picasso: all of the parts that make up the whole are accounted for, but the sense of proportion and placement doesn't fit with reality. The question of how oceans crowded with life actually functioned is still "back-of-the-envelope stuff" he says. While modelling Hecate Strait, for example, researchers turned up evidence that Pacific bluefin tuna once consistently appeared along North America's northwest coast during cyclical warm-current years. The tuna left a lasting impression in indigenous oral histories—not surprisingly, given that tuna-hunting trips could involve tracking the schools by the eerie green light of phosphorescent plankton churned up at the water's surface, then spearing fish that can be three metres long, weigh as much as five men, and swim at speeds of up to eighty kilometres per hour. Bluefin tuna today are known to migrate between waters south of Japan and the coasts of California and Mexico, passing at their nearest perhaps 1,200 kilometres to the south of Hecate Strait. They have not been a part of the Pacific Northwest's fauna in living memory, and their former role is just one of the many mysteries of a bygone natural order.* "We don't know how it worked," says Pitcher. "What we know is that it did work, because it was there."

It is an axiom from the history of nature that the planet is nowhere untouched; even remote Kingman Island, "pristine" in modern terms, is raided from time to time by fishing boats hoping to profit from Asian demand for shark-fin soup, and climate change may have affected its reefs. Yet Kingman Reef has helped reveal how the seemingly impossible bounty of the past was, in fact, possible after all.

When scientists measure the abundance of plants or animals in a given area, they frequently do so by biomass, or the weight of living things. On Kingman Island's reef, an estimated 85 percent of the biomass was accounted for by sharks and other top predators. This defied belief. Biomass is normally expected to

---

* Another shift: The average size of individual animals appears to shrink under hunting pressure in many species. Some shellfish have steadily decreased in size over 10,000 years of human impacts on the Channel Islands off California—red abalone have gone from whoppers that would span a dinner plate to mouthfuls that would have to be served by the dozen. Tiger sharks off North America's east coast have shrunk from a typical length of eight feet to about half that size, and silver seabream have grown smaller by about 50 percent since European arrival in New Zealand. Even the seabream's behaviour appears to have changed. Where once they were frequently seen from boats or shore, and even fished for with rifles or spears in shallow water, they are now considered "spooky," avoiding even divers.

be organized in a basic pyramid shape, with so-called primary producers, such as plants or plankton, forming a wide base layer that feeds a middle level of secondary consumers, often thought of as prey animals, which in turn are hunted by a smaller number of predators—the capstone of the pyramid. Kingman Reef turned this bottom-heavy structure upside down. It was the first time that an inverted biomass pyramid had ever been recorded among a community of species larger than the microscopic.

It took several years of research before the improbable nature of Kingman Reef could be explained. Picture a clock with only a second hand and an hour hand. Inside the clock, a small gear counts off the seconds, its teeth engaging with a much larger gear that counts off the hours. The small gear spins rapidly, but with every rotation it turns the larger gear only a fraction. An ecosystem that is top-heavy with predators works in much the same way. Most of the prey fish move quickly through their life cycles—it's estimated that 99 percent of small reef fish are eaten each year. Still, by hiding in the coral, growing quickly, coming into sexual maturity at a young age, and producing millions of young, enough prey fish survive to maintain their populations. Sharks and other predators, meanwhile, grow slowly, mature late in life, produce few young and live for many years. Somehow, the two have found a balance—one that is far richer than on "normal" reefs affected by fishing, pollution and other human influences. A typical stretch of reef at Kingman is home to about four times the weight of fish as a similar area on nearby Kiritimati atoll (also known as Christmas Island), which has a population of five thousand people. More heavily exploited reefs have fewer fish still.

Kingman Island is not a freak anomaly. Several other reefs—all of them relatively remote from human interference—are now known to be top-heavy with predators. In fact, it's quite possible that this is what many if not most reefs would look like in a pristine condition. Countless sharks appear in the histories of coasts that today are densely populated with people: Accounts from Florida in the 1880s describe sharks "swarming" around fishing wharves, and, as late as 1920, driving fishermen from their grounds by attacking their catch. Sea shanties once warned of the blue and porbeagle sharks that lurked just offshore of the British Isles, and as recently as the 1960s some six thousand sharks a year were being hooked off southwestern England; the same fishery brings in less than 5 percent of that number today. In just the past forty years, the worldwide population of great sharks—those species that grow to more than six feet in length—has likely declined by 90 percent or more. Reef sharks are down by as much as 97 percent.

When Sala and his colleagues first circulated their research from Kingman Island, they met with flat scepticism from their scientific peers. Then they sent out the paper again, this time with photos of the schools of sharks included. The doubters grew quiet. Before long, some were saying they wanted to visit the reef, just to experience it for themselves.

♦ ♦ ♦

Abundance begets abundance. Consider another ecological puzzle, this time involving shrimp-like animals called krill in the Southern Ocean that surrounds Antarctica. Before industrial whaling, baleen whales in the Southern Ocean ate a mass of krill equal to more than half the weight of all the fish caught by people worldwide each year today, with the critical difference that we are severely depleting our fish stocks, while the krill catch by whales was apparently sustainable. When Antarctica's great whales were nearly wiped out in the 1960s, scientists expected krill populations to boom. Instead, they did the opposite, crashing by some 80 percent by the end of the century. Climate change was suspected. Then in 2008, Victor Smetacek, a polar biologist based in Germany, proposed a more intriguing explanation. Perhaps, he argued, whales not only eat krill, they also produce the conditions in which krill can thrive. In turn, larger numbers of krill support a growing population of whales, and both increase on an upward spiral. He called it "the food chain of the giants."

It is sometimes known by a different name: "the whale-shit hypothesis." Here, roughly, is what the theory proposes. While whales feed on krill, krill feed on plankton. In order to bloom, however, plankton need sea water that contains sufficient iron, and the Southern Ocean receives only a small amount of iron from natural sources, such as fallout from the atmosphere or runoff from Antarctic soil.

Whales, though, boost the iron content of the ocean in two important ways. First, they recycle the iron that's available. When krill eat plankton, the krill themselves end up loaded with iron; whales then eat the krill and, later, defecate in liquid form near the surface of the sea. The quantity of iron-rich excrement produced by individual whales is impressive: one research team described whale manure, almost poetically, as "flocculent fecal plumes," each "about the size of our inflatable boat, and the colour of oversteeped green tea." The feces refertilizes the ocean with iron, spinning the nutrient through the food chain multiple times before it inevitably sinks to the sea floor.

Second, whales have the capacity to retrieve the sunken iron. Sperm whales, along with a surprisingly long list of smaller toothed whales, dive to incredible depths. Adult sperm whales routinely dive one kilometre below the surface, and are thought to be capable of diving three times that depth during epic aquatic journeys that can last two hours. (By way of comparison, the deepest human dive without mechanical assistance plunged just one hundred metres into the sea and lasted only four minutes.) Despite being as dependent on air as you or I, sperm whales spend more than 70 percent of their time holding their breath. When diving to terrifying depths, where the water is pitch black and presses in with crushing force, the whales shut down every bodily function not essential to survival. That includes digestion. Whales don't excrete in the deeps. They hunt iron-rich prey such as giant squid down there, but empty their bowels at or near the surface, where the nutrients can once again feed blooms of plankton.

It all sounds like a biologist's April Fool's joke, but the effects are significant. The estimated twelve thousand sperm whales that currently live in the Southern Ocean draw fifty tonnes of iron to the surface every year. Now consider the fact that the estimated pre-whaling population of sperm whales is 120,000—ten times today's numbers. Suddenly, five hundred tonnes of deep-ocean iron is reaching the surface each year, much of it then recycled at the surface by baleen whales feeding on krill. Whaling records from the Southern Ocean indicate a historical population of baleen whales that was so high that biologists couldn't understand how there was enough food for them all in such an iron-poor environment, and some thought the numbers had been over-reported. Instead, the whales may have been boosting the productivity of the entire ocean, making their own extraordinary abundance possible. Blue whales—the largest mammal ever to live on earth—are now thought to have been as much as one hundred times more numerous in the Southern Ocean than they are today.

The idea of fertilizing the oceans with iron has recently inflamed debate in a much different sphere: the fight against climate change. When a kind of plankton known as diatoms bloom, they not only take up iron from the sea, they also remove carbon from the atmosphere—the same carbon that, mainly due to pollution from the burning of fossil fuels, now threatens human civilization with catastrophic global warming. Researchers theorized that when diatoms died, their bodies would sink to the ocean floor, where the carbon they contained would be kept out of the atmosphere for centuries or even millennia. Scientists are now testing whether ocean iron fertilization could produce and sink enough diatoms to combat carbon pollution from human activities. One of the most important studies so far was led by Smetacek himself—he of the "food chain of the giants" hypothesis. In 2012, his team reported that they had successfully used iron seeding to create a plankton bloom, with at least half the carbon that the plankton removed from the atmosphere ultimately ending up at the bottom of the sea.

Ocean iron fertilization is known as "geoengineering," or tinkering with the earth's biggest systems in order to stabilize the climate; other proposals include spreading reflective particles in the atmosphere to bounce the sun's warming rays back into space, and using artificial "super trees" to capture atmospheric carbon for storage in underground chambers. Even among its advocates, geoengineering is widely considered a perilous, last-ditch response by a world community that appears unable to find a way to rapidly reduce its dependence on fossil fuels. Potential risks from iron-seeding experiments, for example, include irruptions of toxic algae or "dead zones" where plankton blooms starve every other species of oxygen. On the other hand, iron fertilization's potential as a tool to fight climate change has made it impossible to ignore. Marine policy experts estimate that the carbon that could potentially be removed from the atmosphere by ocean iron fertilization would be worth $1 billion on the emerging international carbon-trading market, which puts a price on carbon pollution.

In the past, however, large numbers of whales fertilized the ocean with iron, probably with few if any risks. But how much carbon did they really remove from the atmosphere? "To date, the role of whale defecation in influencing carbon export has been overlooked," writes one team of scientists, perhaps stating the obvious. Researchers, however, have begun to crunch the numbers. Calculating the impact of sperm whales alone, they estimate that even today's small population triggers the removal of 240,000 tonnes of carbon from the atmosphere each year. Bring sperm whales back to their pre-whaling numbers, and the amount of carbon removed would reach 2.4 million tonnes—a carbon reduction worth more than $20 million on today's carbon exchange. Sperm whales, meanwhile, are only one of *thirteen* great whale species, nearly all of them decimated during the twentieth century.

Individual whales are themselves remarkable storehouses of carbon. Using conservative estimates of the pre-whaling population, rebuilding just the Southern Ocean population of blue whales would turn 3.3 million tonnes of atmospheric carbon into living bone, muscle and blubber, much the way old-growth trees store carbon in their trunks, limbs and roots over their long lifespans. When compared to the most successful iron-seeding experiments to date, restoring the Southern Ocean's blue whales would be equivalent to two hundred such projects—and that only includes the carbon stored in the whales themselves, without the reductions in atmospheric carbon they catalyze through the iron cycle.

When living things die, their embodied carbon is normally released through decomposition. Most whales, however, sink down—I picture a cradled, rocking motion—through the abyssal depths to the sea floor. Each sunken whale removes its mass of carbon from the atmosphere for hundreds or thousands of years. They once did so in such numbers that whale ear bones, which are the size of a large man's fist, turned up regularly in fishers' nets when they first trawled new sections of sea floor. Whole communities of other species erupt wherever a whale carcass touches down, including at least twenty-eight creatures known only from "whale falls"—meaning that such bonanzas were once common enough that specialized scavengers could successfully evolve. It can take fifty years for a huge carcass to fully decompose, meaning that a whale can "live" after death as long as it did in life.

How much did the ecological abundance of the past shape the atmosphere? No one can say. Climate change ecology is such a new science that the first major workshop on the subject, at Yale University, took place in 2012. But the first glimmers are fascinating. Research on the kelp forests off North America's west coast found a startling difference in carbon storage between those areas where sea otters have recovered and those where they have not. In the restored areas, kelp is one hundred times more plentiful; a totally restored kelp forest would store an amount of carbon worth $90 million on the carbon market today. In another study, maintaining the vast herds of wildebeest on the Serengeti plains of Africa was found to control wildfire so successfully that enough carbon was stored to equal the carbon footprint of every tourist who flies in from around the world to witness the animals. At the very least, according to Yale ecologist Oswald Schmitz, full-force nature was "another wedge, a stabilization wedge" in the face of climate change.

At the most, it was something more. The time period in which atmospheric carbon has rapidly increased is not only the era in which fossil fuel use soared, it's also the era in which the planet's ecology was stripped down and transformed more rapidly and radically than it had been in thousands of years. It is just possible, in other words, that much of the reason we have a new climate is because we have made ourselves a new world.

Fig. 3

HUMAN
NATURE

I ka wa mamua, i ka wa mahope.
(The future is in the past.)
HAWAIIAN PROVERB

Chapter 9.

THE MAKER AND THE MADE

———— • ♦ ♦ • ————

In the 1990s, Mao County in central China ran out of wild bees. The reasons, as usual, turned out to be various. Pesticide use was too heavy, honey hunters were taking too much honey, people were clearing too much of the forests in which the swarms made their homes. But Mao County is an apple-growing region, and in the end not enough bees remained to properly pollinate the trees. By 1997, almost all of the apple growers were doing the job by hand, dipping pollen from blossom to blossom using brushes made from chopsticks and chicken feathers and cigarette filters. That image—human pollination crews teetering in the branches like scarecrows scattered by a windstorm—travelled around the world. It seemed to be an object lesson in the importance of maintaining the diversity of other species, and of how desperate life can become when natural systems collapse.

Fifteen years later, three American researchers published an economic analysis of what they called "the parable of the bees," and turned the story upside down. Mao County's apple growers told interviewers that they actually *preferred* hand pollination. Human pollinators, it turned out, were better at getting to every blossom, performed cross-pollination more efficiently and could work in windy, rainy weather that a bee would never venture out in. What's more, wages paid to orchard workers were often spent in the local area, further bolstering the economy. Worker bees don't head off to the bar or the grocery store when their day is done.

"Destroying and replacing the free gifts of nature can be an economic benefit," the researchers concluded. They might have gone on to argue that we should immediately begin to identify other ecological processes that could be replaced with human labour and technology, but in this case the team, led by the economist John Gowdy of Rensselaer Polytechnic Institute in Troy, New York, moved in the opposite direction. The parable of the bees, they argued, is not that natural systems aren't always valuable, but that it's dangerous to measure the value of nature only in dollars and cents. "Market valuation is an exercise for people who have lost all sense of ecological embeddedness," they wrote. "This is us, the global economic human of the twenty-first century."

Here is the most uncomfortable lesson to be taken from the history of nature: that we can survive—thrive, even—in a degraded natural world. Mao County found it could live without its bees. London and Paris feel much the same way about their long-lost brown bears. Saskatchewan and Kansas prosper without their tumbling buffalo herds; China and Egypt carry on in the absence of their elephants. Large areas of the globe have lost all or nearly all of their largest animals and most ancient forests, and yet they remain desirable locations for people to live. The entire continent of Europe is a tastefully appointed ecological wasteland—rich in human culture, antiquities and innovation, but poor in the abundance and variety of species.

The nature that we live with is a choice. Human beings have witnessed a broad spectrum of states of nature, ranging from continents and oceans teeming with megafauna to landscapes so reduced and transformed that we have to get the neighbours together just to pollinate the flowers. No one, certainly no single generation, decided our trajectory from a richer to a poorer eco-logical world. Different arrangements of people have taken more or less from their environment in different places at different times, but as a general pattern each one gave itself limited permis-sion to degrade nature as they knew it, then adapted to live with the consequences. It's another maxim from historical ecology: we excuse, permit, adapt—and forget. We've been adrift as a species, making choices without remembering what our options are.

After tens of thousands of years, the accumulating effects have become more difficult to ignore or overlook. In the year 2000, Eugene Stoermer, a biologist, and Paul Crutzen, an atmospheric chemist, published a paper arguing that the Holocene—the name given to the epoch in geological time that began at the end of the last ice age—was over, and a new era driven by human transfor-mation of the globe had begun. They called this new epoch the Anthropocene, which translates roughly as the Human Age.* As a starting date, they proposed the late eighteenth century, pointing

* Stoermer had coined the term in the 1980s.

in particular to James Watt's invention of the steam engine, which accelerated humankind's technological power. As evidence that our species' impact was substantial enough to be thought of as a new geological epoch, Crutzen and Stoermer pointed out—among other factors—that the Anthropocene planet is home to 1.4 billion cattle; pumps more sulphur dioxide into the atmosphere through coal and oil burning than all natural sources of that gas combined; features a new and widespread weather condition called "smog"; has had 50 percent of its land surface physically reshaped by human hands; and is so badly polluted with carbon dioxide from fossil fuels that the global climate will likely be affected for at least the next fifty thousand years. We have changed the earth to such an extent that even if it was possible to suddenly lay down our tools, we would still end up with a world of our own creation. The choices going forward are our own, however squeamish we may be about human hubris, however unwilling we may be to shoulder responsibility for the rest of creation. We are, as one wildlife biologist put it, "condemned to art."

This is nowhere more true than when it comes to biodiversity—the oddly technocratic term we have given to the variety of life that inhabits the planet. Attempts to estimate the total number of species have produced figures ranging from 3 million to 100 million; a recent effort—considered conservative, as it is based on the degree of known variety within the most fully studied categories of species—predicts that the earth is home to approximately 8.7 million cell-based life forms, 87 percent of them still unknown to science. Not surprisingly, the species that have already been catalogued tend to be the most conspicuous—relatively large, widely distributed and numerous. The species that have yet to be discovered will mainly be small, local and obscure.

Let us be blunt: we can lose a bunch of them without threatening our own survival. Warnings that we put ourselves in danger when we drive species to extinction often make use of the "rivet hypothesis," first put forward by the biologists Paul and Anne Ehrlich in 1981, which suggests that the components of any ecosystem are like rivets on an airplane: you can only remove so many before a structural failure causes a catastrophic crash. It is probably more accurate to compare nature to a city. You can subtract from a metropolis nearly endlessly over time. Many types of removals, such as the elimination of a single parking sign or the bench at a particular bus stop, will go almost unnoticed. Even major changes—imagine the closure of every sports arena or the permanent loss of mobile phone service—do not make a city unlivable. Daily life for the residents may become more unpleasant and less aesthetic, it may constantly demand innovation and adaptation, but life itself tends to go on with a shrug and a weary smile.

Many of the more than four thousand species that the International Union for Conservation of Nature classifies as critically endangered could wink out of existence without the slightest ripple being felt around the world. Would we miss the Bolivian chinchilla rat? It's known only from a single misty forest of about one hundred square kilometres, located nearly two thousand metres above sea level in the mountains of central South America; even there the rat may be specialized only to rocky areas. *Cola praeacuta*, a small evergreen tree from the same lineage that gave the world the caffeinated cola nut, has no recorded common name in any language and is found only in the foothills around Mount Cameroon in Africa; were it to vanish from the face of the earth, other trees would readily grow in its place. The Himalayan quail, once a fairly common species on the grassy slopes of steep peaks in a patch of north-western India, has not been seen since 1876. It's believed that the quail may still exist, but in terms of its impact on human life, it might as well already be gone.

The disappearance of these species would change the world: among other things, the chinchilla rat helps distribute certain seeds in the cloud forest; *Cola praeacuta* is useful as firewood for local villagers; the Himalayan quail surely shaped the mountain meadows in subtle ways, and once upon a time was hunted as a game bird. Still, all three species and plenty more could shuffle into the void without endangering humanity's future. We might add, more or less at random, the poso bungu, a small fish known from a single lake in Indonesia; the Kythrean sage, which grows on limestone hilltops on the Mediterranean island of Cyprus; and the bizarre-nosed chameleon of the mountains of Madagascar. The list is long. A city has a lot of parking signs, a lot of bus-stop benches.

◆ ◆ ◆

There are limits. We may be able to ride the current wave of extinction, but we do know we can't ride it all the way down. Taken as a whole, natural systems are the basis of life on earth, as easy as that may be to forget in times when meat comes from the grocery store and water from the turn of a tap. Grasslands still feed our livestock, and forests still store and filter our water. Even wild foods remain critically important to modern life: the oceans provide us with about 90 million tonnes of fish each year, feeding billions of people every day. The community of living things, from microbes to megafauna, is endlessly producing oxygen, generating topsoil, stripping our chemical pollutants from the water, slowing erosion, controlling pests, moderating the climate. Economists try to place a value on these "ecosystem services," but the numbers—one pioneering estimate was $33 trillion—are too large to be more than abstractions, and too small to express the reality that we are nowhere close to being able to replace a living world with human technology. Nature is priceless. As the environmental historian Donald Worster says, "We have not learned how to live on a planet that is dead."

The contributions of individual species, too, surround us every day. Thousands of medical drug treatments are derived from plants, animals, fungi and other life forms; most famously, perhaps, childhood leukemia can now often be cured thanks to a drug developed from the rosy periwinkle, a flowering plant from Madagascar that once was threatened with extinction. Add up the most important crops of every nation: we currently rely on 103 species to supply 90 percent of the world's plant-based foods. Our capacity to feed a growing population will surely benefit from the thirty thousand additional species that are believed to have edible parts, including such oddities as the marama bean, *Tylosema esculentum*, an African vine that can produce large edible tubers and nut-like legumes in 50-degree Celsius heat and less rain than falls in Death Valley, California; or the medlar, a Northern Hemisphere fruit that is harvested mid-winter, after it has started to go rotten on the tree. Nor is the practical value of a species always so earnestly survivalist: the first man to deliberately jump out of an airplane and land on the ground without using a parachute, Gary Connery of Britain in 2012, did so in a wingsuit inspired by the anatomy of flying squirrels while using steering techniques he learned by watching birds of prey called kites.

We've begun to understand that even our own bodies are richly diverse ecosystems. Some ten thousand varieties of microbe have now been found on and inside of healthy human beings, in mixtures unique to each individual. These other species' cells outnumber human cells ten to one, though it may be a comfort to hear that these additional cells are generally far smaller, so that less than three kilograms of each person's body is accounted for by them. Most are beneficial, helping to digest food, process vitamins and prevent disease, and many are so thoroughly adapted to living as a part of us that growing them in a lab is as difficult as growing an orange tree in the Arctic. We wouldn't live long without these companion life forms, and vice versa; biodiversity is a part of what makes us human.

♦ ♦ ♦

The idea that other species make important contributions to our lives and, taken as a whole, form the basis for our continued existence on this planet, is one of the most important of the past twenty years. It has also had the unfortunate side effect of encouraging the view that every living thing should be valued in terms of its practical, measurable usefulness to people. Extend this notion far enough, and it even becomes possible to weigh whether the absence of any particular species—such as the bees in Mao County—may be more valuable to human interests than its presence.

Several years ago, I found myself asking a wildlife biologist named Joe Truett to justify his work in exactly such utilitarian terms. We were sitting under an elm tree in a desert grassland preserve in southwestern New Mexico. Bison, which had been brought back onto the landscape, were filing across the sun-burned basin. On the horizon, in the craggy Fra Cristobal Range, mountain lions once again hunted a reintroduced population of desert bighorn, and from our patch of shade Truett and I scanned the skies for the aplomado falcon, a bird of prey with beautiful gunmetal-blue and cinnamon markings that had been released in the area as part of an effort to re-establish a U.S. population extirpated in the 1950s. Why, I asked, did all or any of this matter?

Truett answered the question in what is now the usual way: biodiversity contributes to the stability and resilience of the planet; ecological services have economic value; restored landscapes serve as useful reference points for understanding natural potential and the effects of change. None of these arguments was good enough for me that day. I wanted him to somehow prove that the scene we were looking at was as essential as it felt to me in that moment. I wanted a single crowning reason that we should live with more natural abundance, not less, a richer rather than a poorer state of nature.

At last, Truett said this: "I like it."

It isn't said so plainly nearly often enough: we can simply *prefer* a wilder world. Debates over how to use natural wealth often pit one set of interests against the other: fishermen versus fisheries scientists; oil companies versus indigenous people; housing developers versus outdoor sports enthusiasts. The question at the heart of these conflicts—*whose nature?*—is not the first one we should be asking. That larger question is *which nature?* What kind of nature do we want to live with? The possible answers are not the domain of science or economics alone, but involve every one of us. As a question of individual and collective values, it is perfectly legitimate to like nature, want more of it and to stake your position on grounds of curiosity, awe, mystery and delight.

We won't keep species like the chinchilla rat, *Cola praeacuta*, or Himalayan quail alive because we need to, only if we want to. Every species, when closely examined, has qualities that could one day lead to breakthroughs in human advancement, but in the meantime, they are never less than cause for wonder. The common North American bird called the chickadee, for example, appears to grow a larger brain in autumn, when it needs to remember where it is caching seeds for the winter, then shrinks it again in the spring in order to conserve energy for mating. The great frigatebird is able to put half of its brain to sleep at a time, helping to explain how the animal has been recorded in flight over open ocean for up to twelve days straight. The Eskimo curlew, hunted to extinction in the early twentieth century, is thought to have had among the most efficient muscle fibres of any bird; if a curlew was a small airplane, it could fly 1,500 kilometres while burning the same amount of fuel that ordinary aircraft consume while taxiing out to the runway. Another avian species, Australian brush-turkeys, construct piles of leaves and branches, often as large as five metres in diameter and waist-high to a human being; the birds' eggs are then incubated by the warmth of the decaying vegetation. The compost mound of an experienced male brush-turkey will never vary in its internal temperature by more than two degrees.

The duck-billed platypus, a freshwater mammal that is Australia's answer to the beaver or otter, can apparently "smell" the electricity generated by the movements of prey while it feeds blindly in muddy water. Polar bears were once thought capable of swimming no more than an already impressive 120 kilometres, until a female was recently recorded covering 687 kilometres in an uninterrupted nine-day swim, without sleep, water or food, all while navigating due north using senses that no one yet can explain. Polar bears, incidentally, are so well insulated by their fur coats that they are effectively invisible to infrared cameras. Elephant seals can hold their breath for up to two hours; the human record is eleven minutes.* Ant cities in Africa may involve the removal of forty tonnes or more of soil, plunge eight metres into the ground and have hundreds of well-ventilated under-ground rooms, all built without any guiding architectural mind.

If we are now coming to understand that the hundreds of other species that live on us or in our bodies are a part of what makes us human, then perhaps the same can be said of the species that live outside of us. The planet's other life forms reveal so many ways of being that we could never imagine them if they didn't already exist in reality. In this sense, other species don't only have the capacity to inspire our imaginations, they are a form of imagination. They are the genius of life arrayed against an always uncertain future, and to allow that brilliance to wane out of negligence is to passively embrace the death of our own minds.

◆ ◆ ◆

We have been looking at the natural world as something sepa-rate from humankind, using the common definition of nature

---

* People have held their breath longer (though still nowhere near as long as a seal), but only by first hyperventilating a gas mixture with a much higher oxygen content than is ordinarily found in the atmosphere.

as everything that is not us and is not made by us. It's one useful way to see the world, but to gain a wider view, it is ultimately essential to bring our own species into the picture—just another living creature, after all, as miraculous as the rest. The question—*which nature?*—applies to human nature as well.

Michael Soulé, one of the founders of conservation biology, believes that the nature we surround ourselves with goes beyond questions of values to actually shape human beings as a species. In the postwar era of the twentieth century, anthropologists doing research around the world began to notice a remarkable pattern: hunter-gatherers could see better than the rest of us. In fact, it appeared that the further an individual was separated from hunter-gatherer ancestors, the more likely he or she was to be genetically predisposed to myopia, also known as short-sightedness. The effect was dubbed "relaxed selection"; when a group of people no longer depends on long-distance eyesight to feed themselves or keep watch for menacing animals, the common genetic mutation that causes myopia will have little or no effect on a person's likelihood of surviving or producing offspring. The evolutionary pressure on that trait is lifted. A keen nose, too, was once as vital to human survival as it continues to be for many wild animals. Some 60 percent of the genes associated with the sense of smell are now inactive in most people, a loss that has likely taken place only since the dawn of agriculture some ten thousand years ago.

Soulé argues that relaxed selection may also be weakening our bond with nature. Research into the methods hunter-gatherers use to classify plants—which might, for example, allow them to determine whether a newly encountered berry is closely related to known edible or poisonous varieties—has turned up surprising similarities to the systems applied by professional botanists. The anthropologist Richard K. Nelson, who spent years among the indigenous people of northern Alaska, notes that "the expert Inupiaq hunter possesses as much knowledge as a highly trained scientist in our own society, although the information may be of a different sort." Since the dawn of agriculture, such knowledge has been less and less useful. A predisposition for multi-tasking or an innate capacity to endure the stress of urban life is more important to survival today than the ability to inhabit the mind of a tiger or predict a storm through the behaviour of birds.

It's possible, Soulé says, that this explains why many biologists, including himself, feel so profoundly alienated from mainstream society. The thought came to him over a lifetime—Soulé is in his seventies—of trying to inspire people to care for the natural world. As a scientist, his research helped explain the cascade of impacts that can result when predators are removed from ecological systems, yet hunting and habitat destruction continue to empty the world of predators. He's a long-time Buddhist, but sees no globally influential religious practice that consistently instills an appreciation of nature in its followers. In 1991, Soulé helped establish the Wildlands Network, the first major environmental group to focus on large-scale rewilding; he describes efforts to date as "a failure." It isn't that these various approaches—the scientific, the spiritual, the activist—have accomplished nothing at all, but that in every case those who put the living planet as a whole ahead of short-term human interests remain a small minority.

"Those of us who are biocentric or ecocentric don't understand why people can callously and gratuitously go around destroying nature," Soulé says. "We say, 'How can they do this?' But we're different. We're wired to love different things than other people are."

The degree to which we are or are not genetically inclined to care for nature remains largely unmapped territory. If Soulé is correct, and the ability to empathize with the rest of the living world is now thinly scattered and weakly felt at the level of the human genome, then the future of nature would seem very bleak indeed. A more optimistic possibility is that ecocentric thinkers like Soulé represent the continuation of our innate bond with nature in the same way that piano prodigies embody our capacity to make and enjoy music. We don't say that the rarity of gifted musicians represents the slow fade of rhythm and melody from human culture—the world is still home to a lot of church choirs and kitchen-party guitarists. But music is more accessible than ever, while our relationship to nature is increasingly distant and disconnected. Picture a world in which the history of music is largely forgotten, songs are heard less and less often and musical instruments are poorly understood by the vast majority of people.

When it comes to the natural world, this is us, the global economic human of the twenty-first century. Whether by nature or by nurture, Soulé's warning is the same: when we choose the kind of nature we will live with, we are also choosing the kind of human beings we will be. We shape the world, and it shapes us in return. We are the creator and the created, the maker and the made.

Chapter 10.

THE AGE OF REWILDING

———— · ♦ ♦ ♦ · ————

The twentieth century was the golden age of conservation. When the first calendars of the year 1900 were hung, only a handful of countries on earth—including the United States, Australia, Canada, New Zealand, Mexico and Russia—had national parks. One hundred years later, a United Nations survey found more than 100,000 protected areas in 125 countries, almost all of them established since 1962, the date of the first international parks congress and, incidentally, the year that Rachel Carson published *Silent Spring* and helped catalyze environmentalism as a mass movement.

Similarly, the 1900s began with the first ever multilateral agreement to save species from extinction—a one-year moratorium on sea otter and fur seal hunting that, like so many treaties today, was more impressive on paper than in practice. An informed observer from that time would be surprised to learn that many species that seemed doomed to extinction have instead survived through dedicated human efforts on their behalf. We have the twentieth-century conservationist to thank for the continuing presence, however tenuous, of a long list of species including the plains buffalo, café marron tree, Amur tiger, California condor, Montserrat orchid, blue whale, Saint Helena boxwood, mountain gorilla, the wild horse now known as the takhi,* several species of rhinoceros, the world's smallest water lily, and, yes, the sea otter and fur seal.

During the last half of the past century, however, serious questions began to emerge about the kind of world conservationism was creating with its endless rearguard battles for the planet's last patches of wilderness. In the 1960s, American biologists Robert MacArthur and E. O. Wilson began to study island biogeography, or the science of what lives on islands, and how and why they do so. The two determined that the size of an island and the number of species that live on it are correlated: all else being equal, bigger islands have more species and smaller islands have fewer. This "area effect" was first shown experimentally on small mangrove hummocks off the Atlantic coast of Florida, but later research showed that the phenomenon could also be landlocked—parks, for example, are often islands of wild surrounded by countryside that has been totally converted to human uses. In 1995, biologist William Newmark showed that the number of species in each national park in western North America declines in an almost perfect line according to its size. Almost every protected area Newmark surveyed, among them such globally celebrated wildernesses as Banff and Yosemite, has

---

* Formerly called "Przewalski's horse."

lost not only small and little-known species, but large mammals ranging from caribou to wolves. The world's original national park, Yellowstone, lost its grey wolves in the 1930s, resulting in a cascade of effects. Elk and other grazing and browsing animals boomed, until no young willow or aspen trees were surviving to replace the old ones; the number of large willow and aspen eventually declined by 95 percent. Competition for saplings also drove down the population of beavers, and with them went the rich aquatic habitats they create with their dams. Beginning in 1995, wolves captured in Canada were released in Yellowstone, which is now home to one hundred of the animals divided between ten packs, and the forests and beavers began to recover. Most if not all other parks have been similarly transformed in ways that cannot be reversed without hands-on intervention by human beings.

Some 60 percent of the planet's terrestrial protected areas, meanwhile, are islands of green no more than ten square kilometres in size, amounting to parcels small enough to cross in half an hour on foot. Of those tiny parks, fully one-third cover one square kilometre or less, a span that even a tortoise could cover before lunchtime. In 1996, nature writer David Quammen warned that modern nature preserves could be compared to a Persian rug cut into pieces: the result is not a collection of small but complete rugs, but rather a scattering of tattered fragments.

Conservationism is not dead. Even in its iteration as a protector of last-chance landscapes and endangered species, conservation still has urgent goals to meet. The idea is also evolving: proposals now exist on every continent but Antarctica to reduce the island effect by creating corridors of wild land between parks, preserving spaces large enough for even the farthest-roaming species. Examples include the Yellowstone to Yukon project, which would stretch 3,200 kilometres up the Rocky Mountains from Wyoming to the Yukon Territory; the Great Limpopo Transfrontier Park, which would connect and expand protected areas along five hundred kilometres of border country between South Africa, Zimbabwe and Mozambique; and the European Green Belt, which would turn the former no-man's-lands of the Soviet-era Iron Curtain—a human exclusion zone that became an unanticipated refuge for wildlife—into a continent-spanning network of nature reserves.

But conservation is not and has never been enough. Its most fatal flaw, perhaps, has been to encourage the separation of people from nature: parks here, humans there, and there, and there. To date, none of the continental-scale conservation campaigns can be said to be a success; as the fraction of the earth set aside for the preservation of non-human life approaches 15 percent and bumps up against growing global populations and material wants, resistance to human exclusion will only increase. The struggle to live with a wilder nature is sure to continue, but it won't resemble conservationism as we've known it. It will be different enough that it only seems fair to call it by another name. Ours will be an age of rewilding.

◆ ◆ ◆

The quickest criticism of ecological restoration is that it is a nostalgic science, as foolishly romantic a notion as any other form of time travel. It's true that we can't live with nature today in the same way we did on a planet loosely peopled by hunter-gatherers. It is likewise hopeless to want to roll back the clock on every introduced plant and animal, returning each to its native continent. Picture the scale of the task just on Ascension Island, a British territory in the South Atlantic west of Africa, where 83 percent of the known plants are introduced species. Or, for that matter, Canada, where nearly a quarter of plant species have arrived in the past five hundred years. For the state of New York, the figure approaches one-third. In New Zealand, more than six hundred new plants arrived in a single fifty-year period in the nineteenth century, introduced species were once encouraged on the landscape by "acclimatization societies," and thirty-one species of mammal—all of the islands' furbearing creatures save for seals and bats—are non-native.

But it's also true that nature is not historical in the way that cultural artifacts can be. An elephant is not a horse-and-buggy, made obsolete by technological progress. A hermit crab is not a rotary telephone. A rainforest is not a fashion that can be left behind by new tastes or ideas. Every species still in existence is exactly as contemporary as you or I, and nature's potential—its capacity to sustain abundance and variety—remains unchanged. It is this potential, rather than some replica of the past, that awaits restoration. Nature is still with us, constantly available. We need only to remember, reconnect and rewild: to remember what nature can be; reconnect with it as something meaningful in our lives; and start to remake a wilder world.

In a very few places, it really is possible to revisit something like the prehuman past, and the rarity of such opportunities is argument enough in their favour. A standout example is the Galápagos Islands, positioned directly on the Equator nearly a thousand kilometres off the west coast of South America. Most people are familiar with the Galápagos as the place that inspired the British naturalist Charles Darwin to develop his theory that species evolve through a process of natural selection. It's easy to imagine that Darwin must have had his great insight in some pristine sanctuary, but no. The islands were among the last places on earth to be discovered by human beings, but that discovery had taken place in 1535. Darwin arrived exactly three hundred years later.

The Galápagos are, however, notable as one of the few corners on the globe where the landscape as seen by the earliest human visitors was actually recorded in writing. What captured these observers' attention was the same creature that fascinates us today: gigantic tortoises, or in the Spanish of the times, *galápagos*. The lumbering reptiles dominated the archipelago—even islets too small to have permanent sources of fresh water were crowded with the beasts. "We lay here feeding sometimes on land-turtle, sometimes on sea-turtle, there being plenty of either sort," said the adventurer and buccaneer William Dampier, who in 1684 wrote the first detailed report on the tortoises, "but the land-turtle, as they exceed in sweetness, so do they in number; it is incredible to report how numerous they are." There were so many that even after a century of butchery by pirates, whalers, explorers, navy men and other sailors, ships were still stopping in to bring as many as eight or nine hundred tortoises aboard as fresh meat for their long, slow voyages.

The scale of such historical slaughters shocks us today, but until the advent of canned goods and refrigeration, tortoises were a miracle food. Not only did they contain flesh so vitamin rich that it prevented scurvy, but also large amounts of fat, which could be spread like butter or, when heated, was said to be as light and flavourful as olive oil. The meat was apparently delicious— "after once tasting the Galápagos tortoises, every other animal food fell greatly in our estimation," wrote one visiting sea captain—and it was always fresh; the tortoises could be stacked like barrels and kept alive without food or water, in some cases surviving for more than a year. In the first weeks after capture, tortoises were even a source of fresh water themselves, which could be drained from their bladders, though it was said that the best came from the membrane that enclosed their hearts.

Early accounts of the tortoises are by no means limited to thoughtless killing and cruelty. Sailors came to know the creatures well, and many admired them. The animals' strength was legendary: large males were commonly more than a metre in length, width and height, and weighed nearly two hundred kilograms; one observer described how a tortoise carried two men on its back "and never regarded the weight." Even Darwin gave in to the urge to ride them.

Worldly men admitted feeling fear when they first encountered a truly huge tortoise, while others remarked on their intelligence—on board ships they were easily trained to come when called or to stay in some parts of the boat and not others. Even an apparent capacity for joy was not overlooked—the beasts reportedly showed unmistakable delight in a cool rain shower or visit to a mud wallow. Above all, the tortoises were—and are—a mystery. No one has yet explained exactly how a breeding population ended up on an incredibly remote jumble of volcanic islands that boiled straight up from the South Pacific sea floor and have never been linked by any land bridge to the mainland. The best available answer simply relies on the fact that tortoises live a long time—the first to arrive could have waited a century for a mate—and that while they swim poorly or not at all, they do float. The latter fact was discovered when ship crews that had visited the Galápagos Islands threw their tortoises overboard to clear the decks in preparation for battle; one American navy captain found fifty tortoises floating around his vessel the morning after a skirmish. *

However they managed it, genetic research indicates that large tortoises were present on the Galápagos archipelago for more than a million years before their first encounter with human beings. By the time Darwin stopped in aboard the curiously named HMS *Beagle*, it was already too late to return the Galapágos to their exact prehuman state. Goats, cattle, pigs, dogs, rats and a handful of other animals, as well as some two dozen plants, had already been introduced to the islands, and several unique species of rodent were probably already extinct, along with two of the islands' distinct races of giant tortoise. Darwin himself ate tortoise meat, even noting flavour differences between the subspecies, though his palate doesn't appear to have kindled his thoughts on natural selection. These were inspired first by the islands' four species of mockingbirds, which he concluded must have descended from a single ancestral species; the much richer variety of tortoises played only a supporting role in developing his theory. He recalled that each subspecies had a different shape of shell. The largest of the islands—boot-shaped Isabela—was home to five types of tortoise, most of which foraged at ground level and had simple, dome-shaped shells. On drier islands, the tortoises had a saddle-shaped rise in the forward edge of the carapace, allowing them to stretch their necks high to feed on cactus pads. It was as though in

---

* In 2004, a giant tortoise washed up, alive, on the east coast of Africa; it was a species known from the Aldabra atoll in the Indian Ocean, 740 kilometres away—the first direct evidence that prehistoric tortoises could have floated to remote islands.

the Galápagos some higher power had created a natural labora-
tory in which to reveal the theory of evolution.

Ironically enough, in the years after Darwin's visit, the
Galápagos rapidly underwent evolution in reverse. Oil hunters
began to kill tortoises for nothing but their fat, leaving the flesh to
rot, while collectors of rare species swooped in to hunt down the
survivors. Darwin had noted that feral pigs and goats seemed to
be everywhere, and these invaders rapidly began stripping the
islands' unique ecosystem to its skeleton. The Galápagos tortoises
were finally protected in the early twentieth century, and populations rebuilt in part through captive breeding programs around
the world; today, giant tortoises in the islands are approaching
10 percent of historical numbers. Traditional conservation—
protecting the tortoises and setting aside their habitat—was not
enough, however, to bolster anything like a full recovery for the
animals. On Isabela, for example, the lush subtropical forest on
which the tortoises thrived did not come back; new growth was
being mown down by introduced livestock, especially goats.
First domesticated in the highlands of western Iran, goats are
famously able to eat almost anything and can survive on drought
rations of water. By the end of the twentieth century, the goat
population on the Galápagos Islands had swelled into the hundreds of thousands. The tortoises could not compete. The
islands required rewilding.

Project Isabela, at the time the most ambitious island restoration ever, began in 1998 with a straightforward goal: to kill the
goats. There was no international publicity campaign, no calendar featuring the charismatic tortoises that would be helped.
According to Josh Donlan, an American biologist who was the
chief scientist on the $3 million project, eliminating the island's

goats demanded "sophisticated funders" who could accept the idea of "killing stuff for conservation."

First, 500,000 rounds of ammunition were imported from the United States. Aerial sharpshooters in helicopters eventually put in the equivalent of fifty full days and nights of flying time, killing an average of fifty goats per hour. Mop-up involved hunting dogs, ground troops and "super Judas goats"—sterilized, radio-tagged females that had been drugged into permanent ovulation in order to reel in any surviving billy goats and their herds. By the end of the blitz, the average density of carcasses left behind was fifteen per square kilometre, and the goat population had dropped by more than 99 percent. The meat fed only scavengers and the soil; Project Isabela's leadership had decided that using the goats as a food source for the island's small human population would only encourage people to reintroduce the animals after the massacre.

Goats have been eradicated from at least 120 islands worldwide, and they are not the only targets. Project Isabela also killed 1,200 donkeys. In 2012, twenty-two tonnes of poisoned rat bait was dumped on a pair of Galápagos islands—an unusually large, but otherwise common, campaign to wipe out invasive rats. "Conservationists are turning poisoners," read a headline in Britain's *Guardian* newspaper. On many islands the most damaging introduced species include rabbits and even *Felis catus*—house cats. Clearly, rewilding will make new and controversial demands on the way we see the world.

With goats no longer grazing, Isabela's giant tortoises are once again megafauna, free to do what they did for thousands of millennia: browse foliage, eat and excrete seeds, churn soil. Around them a new landscape, not the same as the one they knew five hundred years ago but surely familiar, right-seeming to a tortoise, is rising. The great beasts can once again take pleasure in a shady waterhole.

♦ ♦ ♦

On the opposite end of the spectrum are those places on earth that have, in essence, never known a form of nature without human beings. This may strike you as impossible: even Africa, where our earliest ancestors evolved several million years ago, was once home to nothing but non-human life. But there are much more recent examples.

At the end of the last ice age, almost all of Canada and the northern United States, as well as northwestern Europe. and Asia, was covered with thick glaciers. By then, human beings with the same mental capacities as our own had been present in Europe south of the ice sheets for thousands of years. In North America, whether the first people arrived on a land bridge from Asia and descended an ice-free corridor in the middle of the continent or sailed down the Pacific Coast, or both, the story is much the same: human beings were witnesses as new land was revealed by the retreating glaciers. In fact, that time appears to be remembered. Many indigenous cultures' origin stories describe landscapes that started out grey, rocky and largely lifeless. An account of the Tlingit people in northern British Columbia and Alaska is especially vivid, describing a journey by canoe down a river that had tunnelled between towering sheets of ice. The passage would have taken as much courage as any ocean exploration in the age of sail or space-race flight into orbit.

In much of northern Europe, Asia and North America, then, it seems certain that people moved onto the landscape almost from the moment it became habitable, and remained there as our modern ecosystems developed. One such case is in the Rocky Mountains in what is today Banff National Park in Alberta. Banff is the third oldest national park in the world, established in 1885. In the year 2000, Canada's National Parks Act made "ecological integrity" the number one priority within the country's parks. Those words have the quietly revolutionary effect of putting nature first, at least in theory; in the U.S. parks system, for comparison, conservation and "human enjoyment" are officially on equal footing. With the shift to a focus on ecological integrity, staff in every park across Canada suddenly had a mission to determine what is "characteristic of its natural region."

There was only one place to go looking, and that was the past. In Banff, researchers combed through more than two hundred archaeological studies, five hundred historical photographs and every first-person explorer's account up to 1872, looking for glimpses into what the Rocky Mountains were like before the fur trade, the railway, the highway, the suburbs, the cellphone towers and all the other accumulated impacts of modernity. What they found was surprising. Twenty-first-century Banff was synonymous with elk—the herds were everywhere, giving visitors exactly the wild plenitude they hoped for from a national park. The research, however, showed that elk were not the most common herbivore in older times. That position was held by bighorn sheep, which were still present in the park, followed by plains bison, which were not. Banff without buffalo was not a "characteristic" state of nature.

At the time of Banff's founding, the last bison were being slaughtered on the Canadian Prairies—in the end, not a single plains bison remained alive in the wild in Canada, and just twenty-three survived in the U.S.* As a wild animal, as opposed to the buffalo living on ranches and therefore influenced by domestication and inbreeding with cattle, bison have been slow to bounce back. Wild plains bison now number about 20,500 continent-wide, a figure that has barely budged since the 1930s, and they roam a fraction of 1 percent of their former range.

Where they stand tall is in myth and memory. Walk through the Banff townsite and you'll see bison everywhere, from the stuffed heads in the museums off Buffalo Street to the bison carpaccio served at the Bison Restaurant in the Bison Courtyard. And though Banff lost its wild bison more than a century ago, the park has maintained a special relationship with the species. Beginning in 1898, Banff was home to a fenced pasture that came to be known as the Buffalo Paddock. Established to give a home to three critically endangered wild buffalo originally from Texas, the pen later housed a mishmash of the last animals salvaged from across North America. Only in the late 1990s was the Buffalo Paddock taken down and the bison auctioned off to new owners—though not before one lone bull had escaped and survived through a winter, helping to show that wild buffalo could live year-round in the park without being fed or cared for by people.

In a sense, bison have been gone from Banff barely more than a decade, which might help explain why opposition to a plan to restore them has been muted. Hoofs could hit the ground before the end of 2013, but the project has never stirred the kind of divisive debate that surrounds the animals in Yellowstone National Park, the species' greatest redoubt in the United States.

---

* Conversely, wood bison were extirpated in the U.S. but survived in Canada's north.

There, public opinion tends to fragment along two lines: the human versus the wild. Inside Yellowstone, bison are a much-loved symbol of the wilderness; when they wander outside the park boundary, they are treated as a potential vector to spread disease to cattle. The buffalo kills that control the Yellowstone herds when they leave the park—in some years, more than a thousand bison are shot or slaughtered—play out at times like skirmish warfare, with government agents cloaked in secrecy while animal-rights activists run guerrilla interference.

"Yellowstone is still caught up in this idea of what is natural," says Cliff White, a former Banff park warden who is now a consultant and irrepressible advocate for bison restoration. "But you have these compromises you have to make because the modern landscape is not the landscape of ten thousand years ago."

In Yellowstone, despite being home to wolves, cougars and grizzly bears, bison are easily the most dangerous animal. Get gored by a bison, and you can end up with a wound as wide as a cleaver and deep as a chef's knife, all while getting tossed so high in the air that many of the worst bison-attack injuries come from, as one medical researcher delicately puts it, "encountering sudden deceleration on impact with the ground."* In Banff, even the potential safety issues have not shaken the locals, and support for the bison reintroduction is widespread, even among ranchers. Most people who live and play in the park and along its boundaries have already come to accept that mountain living has its hazards. "Almost everyone here has had the bejesus

---

* Reports of deaths and injuries caused by buffalo in Yellowstone typically benefit from context. For example, the first person to die in such an attack was allegedly attempting to stand next to a wild bison for a photograph.

scared out of them by grizzlies, but they still want them in the valley," one lifetime resident told me. Bison, he said, will be "just another gnarly species in the environment." The question of living with a wilder nature may have less to do with risks and challenges than with the degree to which people identify with the idea of wildness. Rewilding is a matter of nature, but also of culture. In Banff, people *want* bison; the presence of the animals fits with the locals' understanding of themselves and the reasons they live where they do. One bison advocate described the restoration effort as "less an environmental campaign and more of a social movement."

That's the simple telling of the Banff bison story—a classic tale of the wild despoiled by humankind, and an opportunity to heal the wounds. There's another version, though, that asks us to go deeper in the relationship between man and beast. When the Banff researchers combed through the bone digs and explorers' journals, they not only turned up definitive evidence of the past presence of buffalo, they also found that hoofed animals in general were nowhere near as numerous as they would have expected. The most likely explanation for this historical paucity of wildlife proved to be the abundance of one other species: *Homo sapiens*. In the centuries before Columbus discovered his New World, the iconic hanging valleys that run between Banff's lofty peaks were effectively walled-in corridors for native hunters. Indigenous people not only hunted the region's big mammals to low numbers, but also took steps to manage those populations, such as regularly burning the land to create better pasture. Having studied the layout of pit houses at some archaeological sites, White believes that native people may even have herded bison into the mountains from the plains, then blocked them into narrow canyons to be killed off as needed—a form of ranching long before the first cowboy.

In other words, it's possible that bison only established themselves in the Canadian Rockies because the mountains' human inhabitants invited them to the party. On the other hand, it's equally possible that the presence of people prevented bison from freely inhabiting the landscape and from spreading across the Rockies into eastern British Columbia and beyond. What is fair to say, at least, is that when it comes to Banff, *Bison bison* and *Homo sapiens* have been bound together for millennia.

As a species, bison represent the human-animal relationship as well as any wild animal on earth. According to Valerius Geist, a specialist in ice-age mammals and professor emeritus with the University of Calgary, the buffalo as a species is in many ways a *product* of human influence. When people first arrived in the Americas, bison were giant, long-horned beasts that stood their ground against predators—easy pickings for hunters with spears and bows and arrows. To survive into our era, the animal evolved to be smaller, warier and fleeter of foot.

Such realizations can be hard to stomach—the old ideal of wilderness as a place untouched by our hands has been one of the most treasured notions of the past 150 years, and helped put a stop to the genocidal scale of hunting in the eighteenth and early nineteenth centuries that threatened the existence of every living thing large enough to shoot, trap or catch on a fish hook. Without the wilderness ideal, Banff might never have been founded.

There are advantages, though, to seeing the line between the wild and human as blurry—it can make it an easier line to cross. If bison are brought back to Banff, the first herd will be human-manipulated in almost every way. They will be the descendants of another captive herd, and fences will limit their movements. Where the fences aren't enough—"you can herd a bison anywhere it wants to go," as the saying goes—a suite of other measures will have to be considered, ranging from hazing the herds on horseback to culling. If Canadians can learn to live with wild bison in Banff, White says, then they might be prepared to try doing so in other places, even outside of parks. The first Banff bison herd will number in the tens, but given time, White pictures them up and down the Canadian Rockies, and possibly beyond, onto the plains themselves. It will be the rewilding not only of a vanished species, but of something lost in ourselves. Only then might we finally be able to say that the wild plains bison has begun to recover.

♦ ♦ ♦

In 2010, workers restoring Scotland's Rosslyn Chapel discovered a beehive built into the stonework of the rooftop pinnacles. A beehive might seem ordinary enough, but there was a critical difference between the Rosslyn hive and most others: it was not designed to gather honey. Instead, it was apparently made simply to allow bees and churchgoers to live alongside one another, for reasons yet unknown. Though the hidden beehive had long been forgotten by its congregation, humans and bees have lived together at Rosslyn for more than five hundred years.

Scientists acknowledge that much of what will surround us in the future will be "novel ecosystems"—new arrangements of species and natural systems. In places, historical context will have little relevance: former species will have been driven extinct, new species introduced, the landscape dramatically altered and even the climate significantly changed. Newness is a modern talisman, but in ecology it is not the most desirable condition. Another maxim: that which is old has proven itself, and that which is very old may contain wisdom. It's anyone's guess how untried forms of nature will function. Nonetheless, a majority of people on earth already live in novel environments: cities. And while many people in urban areas debate the consequences of issues like rainforest logging and species extinction, the ecological losses that occur when a city replaces a natural landscape are, in the words of University of Nevada conservation biologist Dennis Murphy, "well documented and inarguably immense."

The Rosslyn hive is an example of what we might call habitecture—the integration of habitat for other species into structures designed for human purposes. We live with habitecture all the time, though it's almost always unintended. In a few cases—think barn swallows and barn owls—species turn out to adapt brilliantly to the things we build.* More typically, the animals among us are unwelcome stowaways, from skunks that hole up under backyard decks to pigeons that roost, ever so cannily, on anti-pigeon devices. With rare exceptions, wildlife-friendly structures continue to be accidental. Perhaps the greatest example is the Congress Avenue Bridge in Austin, Texas, where crowds gather every summer sunset to watch Mexican free-tailed bats burst forth from beneath the deck—"an umbrella is a good idea," advises a local tourist guide. The bats chose Austin, not the other way around. In fact, the organization Bat Conservation International has roots in an early-1980s battle to persuade

---

* Barn owls are now in decline as older barns make way for new, impermeable outbuildings and small farms with groves of trees for nesting are lost to the huge, treeless fields of industrial agriculture.

Austinites not to exterminate what has since become the world's largest urban bat colony, with more than a million animals.

Rosslyn's hive represents the key principles that could make habitecture possible: it was designed to allow beasts and people to live separately but together, to keep the animals from destroying the structure, and to function without human interference. We might want to add a fourth principle: Rosslyn's beehive happens to be beautiful, with the bees coming and going through holes in the centres of graven stone flowers. Britain, that island of eccentricity, is also home to other models of early habitecture. "Bee boles," or niches built into outside walls to shelter honey-producing hives, may date back in human history as far as 2000 BC; nearly 1,500 are still known to exist, mainly unused, in the United Kingdom. Dovecotes, often built directly into the walls or gables of houses, including those of the wealthy, housed pigeons that were used for meat, eggs and messaging services. Pigeon droppings, which today infuriate condo dwellers and tattoo the foreheads of forgotten politicians immortalized in bronze, were prized as fertilizer. Beyond dovecotes there were deercotes, elaborate shelters for wild deer, while Culzean Castle in Scotland includes a house for wild ducks along with customized tidal pools. There's even a temple for turtles that dates from the 1820s just south of London.

The most familiar modern habitecture is the birdhouse, though the concept has been extended. Today there are woodpecker houses, owl houses and tree duck houses, and also bat houses, butterfly houses and squirrel houses. England has hedgehog houses. The people of Poland traditionally revered the white stork and in the past mounted old cartwheels on their roofs as nesting platforms; today, ready-made stork kits are available. In a strangely heartening gesture, the official Polish government website refers to the country as a "stork superpower."

That habitecture has its limitations is clear. Many species, from the common loon to the Iberian lynx, share space with human beings reluctantly or not at all. Go ahead and construct a perfectly appointed wolverine den for the walls of your sub-urban rancher—if you build it, they will not come. No matter how inviting we make an urban landscape, it will not recreate a living world that is the equal to what was erased to make way for the city. We're left with possibilities that are far from such ideals—visions of nature that are incomplete, engineered, reg-ulated and in which our hands are muddy indeed. So why rewild the metropolis? Why not let cities and towns be wholly human places with whatever crows, rats and cockroaches are willing to eke out a living between the bright lights?

The answer may be that wild animals and plants don't need to live among us so much as we need to live among them. Many people understand this when it comes to the diversity of human beings: it's one thing to ogle exotic tribespeople in a vintage copy of *National Geographic*, and quite another to live in a neighbourhood of Somali immigrants. Making daily contact with even one other culture deepens our ability to appreciate cultures as a whole. Yet the idea of putting out the welcome mat to other species remains a curiously radical concept, so much so that it is artists, not architects or city planners, who are leading the way.

"Symbolically it's quite an important step if we can deliber-ately cohabit with other species," says American artist Adam Kuby, who has launched projects ranging from shields to protect urban amphibians from damaging ultraviolet rays, to public sculptures that incorporate nesting cavities. Kuby might best be described as a habitat artist. His inspiration did not come, as you might expect, from earlier artists such as Friedensreich Hundertwasser, the Austrian wildman whose blending of the built and natural worlds included "tree tenants" that lived in high-rise apartments. Instead, Kuby had mentors of a different order: two birds that flew in through a transom window of his art school and built a nest atop the incubating warmth of a fluorescent light fixture. The birds were ready to integrate with us, Kuby realized—why not the other way around?

He's a part of a small art movement, but at least it is a movement. In England, the London-based artist Gitta Gschwendtner's fifty-metre-long *Animal Wall*, which matched the development of one thousand new apartments and houses for people in Cardiff, Wales, with one thousand nest boxes meant for such birds as pied wagtails, sparrows, starlings,* blue and grey tits, and even bats, is slowly filling with tenants. There are signs of change in more prosaic architecture as well; the free-tailed bats beneath the bridge in Austin have inspired a program to educate highway engineers and managers about how they can accommodate the thirty-seven U.S. bat species, including four that are listed as endangered, that are believed to roost in bridges if given the chance. Banff National Park, meanwhile, leads the world in underpasses and overpasses designed to allow wildlife to safely cross highways. Grizzly bears at first found them confounding; now they use them nearly every day.

---

* Starlings are an invasive species in many places around the world, including North America, but are native to Great Britain.

"We humans, we label things—this is a building, this is a cliff, this is a tree," says Kuby. "Animals don't see the world that way." His work has made him more aware that other species will go to astonishing lengths to live among us, such as a pair of birds he saw flying in and out of a hole in the traffic light that hangs over the intersection of 5th Avenue and 42nd Street in Manhattan, two blocks away from the glitz of Times Square. "The little birds just find these openings," he says.

Among Kuby's dream projects—artworks no one yet has come to him to build—is something he calls *Cliff Dwelling*. It will involve building a niche into a skyscraper, then filling the alcove with an artificial rock roost suitable for peregrine falcons. The birds—which were long an endangered species after proving susceptible to pesticides such as DDT, but have since recovered to a stable and growing population—have been known to nest on inner-city high-rises in places like Chicago and Toronto. Still, a skyscraper has yet to be designed with falcons specifically in mind.

Cliff Dwelling would be a challenge for any community, from its unusual architecture to the sight of falcons killing pigeons in city squares. But it would act as a daily reminder for urban dwellers of the rest of life on earth, of natural cycles and of the fact that we do not need only to rewild nature, but human nature, as well. In Kuby's design, the birds would be visible through one-way glass from within the building. It would be, in other words, the opposite of a zoo. "Here the people are confined," says Kuby. "The falcons are free."

Chapter 11.

DOUBLE DISAPPEARANCE

———·◆◆◆·———

The Hawaiian Hall of the Bishop Museum in Honolulu is three storeys tall and filled with enchantment. The huge room is deep-shadowed, as if the light was cast by torches, and everywhere you look are the scowling faces of totems, the silhouettes of sharks, the blades of daggers carved from the bills of swordfish. More unforeseen than any of these, however, is the lingering imprint of many, many birds.

Behold the royal cloak of Kamehameha I, the political and military genius who united the Hawaiian Islands under one leader for the first time in history just as the colonial era was beginning in the late 1700s. The king's floor-length cape is composed almost entirely of feathers, mainly from the Hawai'i mamo, a sparrow-sized, mostly black bird that happened to have patches of feathers on its wings and rump in the Hawaiian royal colour, yellow—in this case, the brilliant, golden yellow of the heart of a dandelion flower. Kamehameha's cloak still shimmers more than two centuries after it was completed, and some of the feathers are older than that, handed down from at least eight other Hawaiian kings, who in turn had gathered feathers from their own ancestors and defeated rivals. All told, the royal garment is thought to include feathers from sixty thousand or more individual mamos.

Feathered objects abound in the Bishop Museum, from capes to skirts to decorative helmets. Before it became the fiftieth U.S. state, before it was even settled by Polynesian wanderers, Hawaii was a kingdom of birds.* Its waters would have been thick with toothy sharks, jacks and other predatory fish, along with whales, dolphins and a species of monk seal—animals so perfectly streamlined for the sea that their ears are internal and their nipples retractable—found nowhere else in the world. Yet only two species of land mammal, both of them bats, ever made their way to the islands ahead of human settlement, leaving birds to fill every possible niche. Completely isolated, more than 90 percent of Hawaii's bird species ultimately evolved into forms unique to the island chain. Many lost the ability to fly, untroubled by the predators that stalked much of the rest of the planet.

Not only was the mamo found only on the Hawaiian archipelago, but it was also limited to its largest island, Hawai'i, often referred to today as the "Big Island." Mamos fed on nectar, mainly from the depths of vase-shaped lobelia flowers that, like many of the plants on the Hawaiian Islands, had evolved into

* It's conventional among many Hawaiians to use "Hawaii" to refer to the U.S. state as a whole, and the Polynesian "Hawai'i" to identify the largest island of the archipelago.

forms as distinctive as the birds. As a result, the mamo's bill was long and downcurved, as though it was made of wax and its owner had passed too close to a flame. The song of the mamo is remembered as a plaintive whistle—the description brings to mind the traditional Hawaiian saying, "I fly away, leaving disappointment behind." The call was never recorded, will never be known. The Hawai'i mamo is extinct. Even its skins and taxidermies are rare; only eleven specimens are known to exist in the world's museum collections.

Hawaii is the planet's most remote island chain, more than three thousand kilometres from the nearest habitable landfall, and the first human settlers may have arrived there as late as AD 1250—less than eight hundred years ago. These first Polynesian seafarers, perhaps a hundred or so people sailing double-hulled canoes and navigating by sophisticated under-standings of the stars, sun, waves, wind and behaviour of sea animals, carried with them a complete toolkit for survival. Their canoes were the arks of their culture, and any new-found land became a home to their pigs, dogs, coconut palms, banana trees, taro fields, medicinal plants—in all, more than forty species are known to have been brought to the islands by early Polynesians.

The influx made an immediate impression. Studies of bones in ancient layers of sediment show that sixty species of bird, many of them large, flightless and presumably tasty, soon went extinct. Over the same time period, other desirable food spe-cies, such as sea turtles and the monk seal, suffered population crashes. Human hunting wasn't solely to blame; the Polynesians' pigs, dogs and stowaway rats are also thought to have ravaged birds and their nests, spooked seals from their resting places and breeding grounds, and devoured the seeds of native plants and trees. The lowland forests of the islands were quickly transformed, with many native species eclipsed by so-called "canoe plants." The composition of the original forests is so thoroughly lost in time that no one is certain what they might have looked like. Some evidence suggests that trees and plants now known only from Hawaii's high mountains were once found at the shoreline. The uplands may not be these species' true habitats so much as their last refuges.

It was the usual pattern of human history: we came, we saw, we left a deep scar. "Even modest human impact has a pretty significant ecological signal," says Alan Friedlander, a marine biologist with the University of Hawaii. For a century or two after the islands' discovery, Polynesian sailors may have continued their daring journeys back and forth between the archipelago and other South Pacific islands. Then, for reasons no one has yet been able to explain, they stopped. For at least five hundred years, the Hawaiians lived in total isolation. They had no trade with the outside world, no other place to turn for the necessities of life; any islander could, with a walk to the nearest hilltop, see the limits of Hawaii's natural wealth. For all intents and purposes, they were living on a tiny, fragile planet surrounded by outer space.

During that long period of seclusion, something special began to happen. The rate of species extinctions appears to have slowed dramatically. The coral reefs, overfished in the first centuries after the Polynesians' arrival, entered an era in which they declined little, if at all. Some creatures, such as sea turtles, appear to have partially recovered from their earlier depletion, and large areas were left only lightly touched by human hands. Incredibly, the human population increased at the same time, to at least 400,000 people and perhaps as many as 800,000—not far off the 1.4 million inhabitants found in Hawaii today.

Even the birds endured, despite the fact that their feathers became symbols of status and prestige. The Hawai'i mamo had the most desirable feathers of all, and yet the little birds remained plentiful; they would vanish only in the early twentieth century after the arrival of the first Europeans to the islands. In this case, what's most compelling about the species is not its ultimate extinction but rather its long survival. The lesson of the mamo lies not in what went wrong, but what went right.

♦ ♦ ♦

Twenty years ago, Ray Rogers, a Canadian environmental philosopher and one-time commercial fisherman, turned his thoughts to extinction and extirpation. In many cases, he realized, the loss of a plant or animal also marks the end of a human relationship to that species. As bears faded across Europe, for example, so did the festival of Chandelours—the word translates as "bearsong"—that celebrated the end of the animals' winter hibernation in early February. Similarly, as wildlife populations vanished in the nineteenth and twentieth centuries, so did the "market hunting" profession, along with such wild foods as brant goose, diamondback terrapin, bison tongue and Olympia oysters, each of which was once common on dinner tables and restaurant menus in North America. Rogers described each broken link between people and nature as a "double disappearance," a form of environmental amnesia that went beyond mere memory to hollow out our sense of community with the rest of the living planet.* We were losing species from our social networks.

For many people today, the idea of having a social relationship with nature is distant enough to sound ridiculous, as though you could invite a sea lion to a dinner party or exchange email with an ostrich. Our personal connection to—or disconnection from—non-human life, however, continues to shape the world. In 2010, John Waldman, a biologist at Queens College in New York, argued that the decline of spawning fish runs on the U.S. Eastern Seaboard was not only an environmental crisis, but an ongoing social breakdown. The region's major fish species, such as eels, salmon, shad and the "river herring" called alewives, were once astonishingly plentiful, and people of that time created traditions like the "shad bake," gathered to watch salmon leap up waterfalls and recognized the natural abundance in place names such as Sturgeon Pool in New York. As dams and overfishing tilted the fish into decline, the people who depended on them for employment or food, or who simply enjoyed their presence, often did protest. But the double disappearance was underway. Fishermen found other jobs and largely forgot about the fish, while communities soon were feeding themselves with agriculture or imports from

---

* Rogers published his "double disappearance" concept in 1994, a year ahead of Daniel Pauly's "shifting baseline syndrome" or Peter Kahn and Batya Friedman's "environmental amnesia." In fact, the idea that we forget the natural world of the past appears to be regularly rediscovered; here is biologist Raymond Dasmann in 1989: "But one adjusts to slow, deleterious changes in the environment and begins to accept them as normal. Young people, growing up in smog, have no basis for believing that things were better in the past, and could be better in the future if certain actions were taken. The abnormal is accepted as normal and becomes the standard by which future change is measured." An optimist would say this is an idea whose time has finally come; a pessimist might suggest that we are doomed to forget even what we remember we've forgotten.

faraway places. In the end, the absence of fish made the health of streams and rivers less important, and they were put to other uses, for example as dumping grounds for sewage and toxic waste. Today, East Coast shad bakes no longer serve shad, only a few hundred wild Atlantic salmon remain in the United States, and there are no sturgeon in Sturgeon Pool. In fact, of the seaboard's two sturgeon species, the shortnose sturgeon is endangered and the fishery for Atlantic sturgeon is expected to remain closed until at least 2040. Most other fish stocks in the region have declined by 90 percent or more.

"As species disappear," Waldman says, "they lose both relevance to a society and the constituency to champion their revival, further hastening their decline. We need to rewind important historical connections."

Hawaii is one of the best places on earth to carry out that rewinding: a microcosm where culture and nature were completely intertwined. In the centuries ahead of the first visit by European explorers, the Hawaiian islands were monarchies, ruled by royal families who divided the land first into regions, and then again into communities called *ahupua'a*. (AH-hoo-poo-ah-ah. The apostrophe marks a glottal stop, or catch in the throat, like the sound represented by the hyphen in the phrase *uh-oh*.) Ahupua'a were typically wedge-shaped sections of land that ran from the uplands to the sea—"from *mauka* to *makai*," as Hawaiians put it today. In many cases they enclosed the entire drainage, or watershed, of a stream or freshwater spring, but their boundaries depended most on the need to provide each community with the resources to sustain itself independently. For the Hawaiian living in an ahupua'a, care for the land could not have been more obviously linked to survival—misuse your land or waters, and you could not assume that your neighbours would be able to come to your aid.

"If you did something wrong here, the whole system would feel it really quickly, whereas in a continental system, if you do something wrong, well, it'll be a long while—probably generations—before impacts are actually felt and recognized," says Kawika Winter, the director of the Limahuli Garden and Preserve, a branch of the U.S. National Tropical Botanical Garden on the north shore of the island of Kaua'i. "The things that our ancestors figured out are applicable in whatever watershed on the planet you go to, whether it's the Mekong Delta or the Nile or the Mississippi. It's all still a watershed, it's all the same issues."

The Limahuli Valley is one of the few historical ahupua'a that has not been subdivided. It's a spectacular landscape, with a tumbling creek that spills into the valley as a 250-metre-high waterfall and verdant forests that climb the convolutions of volcanic ridges and spires. At the foot of the valley, looming above the coastal plain, stands Makana, a mountain of such primeval presence that it was made an icon of all things exotic as the fictional island of Bali Ha'i in the classic film *South Pacific*. Just over three kilometres from its uplands to the shore and not much larger than New York's Central Park, Limahuli was once home to as many as two thousand people.

Traditionally, the highest reaches of the valley, hidden in mountain folds, were the *wao akua*, the realm of the gods. It was a locus of spiritual intensity, entered almost exclusively by bird hunters—historical photos remember them as wild-looking men with bushy afros and far-seeing eyes—in search of precious feathers. Strict rules applied to these sacred woodlands; in some cases, cutting down a single ancient *'ohi'a lehua* tree, which might stand thirty metres tall, would require a human sacrifice. People had more freedom in the forests below the waterfall, where they gathered products ranging from hardwoods—the islands had no sources of metals—to ceremonial flowers to edible and medicinal plants. The village was built mainly on rocky ground, saving the richest soil for crops. Farther below, where the Limahuli Stream reaches the sea, the Native Hawaiians built and managed fish ponds. Even that was not the end of the ahupua'a—every reef and coastal fishing ground was assigned to a community.

The bird hunters in the wao akua have been replaced by bird biologists today, but you can still find traditional gatherers in Limahuli, as well as terraced fields and fish ponds. The ahupua'a has proved to be a system that makes ecological sense. The protected upland forests harbour biodiversity, prevent soil erosion, and suck up water like a sponge to be slowly released even in dry weather. Lower down, canals known as 'auwai divert up to half of the creek's water to irrigate the crops and then return the remaining flow to the natural channel, which is one of only two streams on Kaua'i that has retained all five native species of freshwater fish.* During the island's some-times savage rainstorms, mountain runoff from less well cared for land turns Kaua'i's coast a muddy red-brown, threatening to choke out life on the reefs. In the Limahuli Valley, the forests hold the soil, and silt that does reach the creek settles first on the croplands, fertilizing them, and later in the fish ponds,

---

* The fish are gobies, and four out of the five species have developed suction cups on their bellies that they use, along with their mouths, to hold their own against the steep creek's rushing water or even to climb waterfalls.

where it triggers algae blooms that help feed plant-eating mullet and milkfish. By the time the stream reaches the sea, its water is clearer than it would be if people were not present.

"Humans are a part of ecosystems—that's the approach we take," says Winter. "When you talk about saving ecosystems, humans are a part of that equation."

◆ ◆ ◆

The idea that people are a part of nature is often raised thoughtlessly, as though the fact that we are carbon-based life forms means we have the same moral culpability as panda bears or redwood trees, and any harm we might do is simply "natural." In the Native Hawaiian tradition, the concept is considerably more demanding. According to the *Kumulipo* creation story—the Hawaiian equivalent of the biblical book of Genesis—the first living thing is a coral polyp, a minuscule life form that is the basic building block of reefs;* from there, life gradually branches out to include human beings, a world view that accords neatly with the theory of evolution. Rather than proclaim humans the pinnacle of nature's progress, however, Hawaiian cosmogony holds that we are new arrivals among respected elders. For example, the taro plant, known as *kalo* on the islands, is specifically identified as the Hawaiians' immediate older brother; humans are called upon to care for the taro, which has its own obligation to keep its younger sibling alive. The starchy taro-root paste called *poi* remains the defining dish of Hawaiian cuisine.

Before Europeans reached the archipelago, Hawaii was home to four hundred types of taro plant. Today, about seventy

---

* In the Bible, the first living thing created is grass.

historical varieties of taro are known to remain; almost all of them can be found in Limahuli's terraced fields. Perhaps it was this wealth of variety, which allowed the Hawaiians to produce food in conditions ranging from floods to drought, that encouraged their remarkable appreciation for biodiversity. No one can say. What is clear is that Hawaiians valued an incredible assortment of wild plants and animals, some as embodiments of godly power, others for their role in ritual or crafts, some as food, still others simply for their beauty. The relationship was one of give and take—it was *social*. The various species gave of themselves, and in return had a constituency of human allies who advocated not only for their survival, but their abundance. The Hawaiian term *kuleana*, which has a meaning suspended somewhere between "responsibility" and "privilege," captures this interconnection. Imagine that a friend calls on you at a time of need. You are obligated to respond, but the obligation is also a point of pride. Your friend asked *you*, placed trust in *you*. It is your kuleana to answer the call—your duty and your honour.

The aspect of traditional Hawaiian culture that most challenges modern sensibilities is the system of *kapu*, or taboos, that governed everyday life and could carry an immediate death penalty if broken. By the time Europeans were writing about Hawaii, it seems clear that kapu was experienced by ordinary Hawaiians in large part as tyranny, and many taboos were quickly abandoned when the system was abolished by the Hawaiian royal family in 1819—the first rule to be broken was the prohibition against men and women eating together. Yet kapu had also played a critical role in sustaining the island's natural richness. It may sound shocking that a person could be killed for such crimes as fishing out of season or bathing in a pool of drinking water, but to ancient Hawaiians, such rules might have seemed like common sense. They would no more threaten the basic conditions that preserved their survival than people today would choose to drive against the flow of freeway traffic.

Consider the story of Hua, the long-ago king of eastern Mau'i. The seven-hundred-year-old legend is considered, as King David Kalakaua wrote in 1888, "one of the most terrible visitations of the wrath of the gods anywhere brought down by Hawaiian tradition." The story has taken on various forms, but I will tell it as it was told to me by Sam 'Ohu Gon III, a Honolulu-based conservationist who has completed both a PhD in animal behaviour and a rite of passage in traditional Hawaiian cultural protocol. We spoke at an outdoor table in the parking lot of a shopping mall.

Hua is a decadent and wasteful king, and he despises the restraint and aloofness of his high priest. Knowing there is a kapu against hunting the gull-like birds called petrels at the shore, where they are easily found and killed, Hua orders his bird catchers to bring him petrels from the mountains. But his hunters are lazy, and they snare their petrels on the coast and rub cinder dust into the feathers to make it appear the birds have come from the slopes of volcanoes. The hunters bring their catch first to the high priest, who suspects the ruse; he cuts open a bird and, finding seaweed and fish in its belly, confiscates the forbidden wildfowl. Knowing they have broken kapu and will be put to death, the hunters go to Hua and accuse the priest of keeping for himself the birds they had killed for the king. Seeing a chance to bring down his adversary, Hua has the priest put to death for treason. Immediately, a long drought begins. The people begin to die, and Hua himself is finally

forced to flee his kingdom. But the drought follows him wherever he goes, and at last he falls dead with no one left to bury
him. He's remembered in a Hawaiian proverb: *The bones of
Hua rattle in the sun.*

"Everybody paid for it," says Gon, leaning back in his plastic
chair. "Not just the bird catchers who tried to get away with it—
all the people die. The priest dies. The king dies. This idea of
kapu being so serious that everybody dies for it really speaks to
how fundamentally important it was to obey those kinds of rules."

Native Hawaiian culture was disrupted and suppressed after
the islands' colonization by Europe and later the United States,
and the ahupua'a network was shattered as properties were
divided among private, mainly foreign, owners. Today, the traditional land ethic endures in only a handful of places, and the
land itself—deeply transformed over the past two centuries—
is often the only code to decipher ancient ways of thought.
Kawika Winter admits that he sometimes struggles to see the
Limahuli Valley through the eyes of his ancestors. It's an old
saying, for example, that the health of the sea depends on the
health of the mountains. The reason is obvious: water runs
downhill to the shore. But it's also said that the health of the
mountains depends on the health of the ocean. Winter could
make no sense of this until 2006, when two endangered seabirds, the Hawaiian petrel and Newell's shearwater, were found
breeding high in the Limahuli Valley. Today, Hawaii's most
important nesting sites are on offshore islands or behind fences
that protect the birds and their eggs against introduced rats,
while overfishing has drastically reduced the seabirds' food
supply. The hidden nests in Limahuli were a reminder that in
the past huge colonies of petrels, shearwaters, albatrosses,

frigatebirds, tropicbirds and boobies filled the cliffs and dug burrows in the highlands, their feces painting the landscape white with nutrients from the ocean.

"That's the way our ancestors viewed everything: How do we engage and interact with this system and make it pump?" says Winter. "It's a different mindset. Every time we lose a tree, a vine, a bush, a little bird, that's a word and a name that drops out of our lexicon. That's a story that we can no longer tell."

♦ ♦ ♦

Hawaii's seclusion ended on January 18, 1778, when Captain James Cook of the British navy, commanding the ships *Resolution* and *Discovery*, stumbled upon the islands while en route to search for the Northwest Passage. Cook encountered a Hawaiian culture in some ways remarkably like his own: deeply religious, ruled by kings and queens and fed by a working class of farmers and fishers. The Hawaiians certainly were not eking out a desperate existence. They showed no interest in trading for the Englishmen's food, instead seeking practical items such as metal nails and high-status goods such as beads; King Kamehameha apparently exchanged a feather cloak for nine iron daggers. Many Hawaiians enjoyed enough leisure time to celebrate human beauty, study navigation, prepare for war against rival factions, practise arts such as dance and carving, or dedicate their time to arranging flowers into the ceremonial garlands known as *lei*. Others practised thrill-seeking sports such as surfing, cliff diving and *holua*, in which wooden sleds were ridden hurtling down mountainsides on beds of leaves. Even today, a visitor to Hawaii may notice Old World similarities—the nostalgic fondness for monarchy among some

Native Hawaiians, for example, or the deep love for a landscape that has few traces of its original nature.

The changes wrought by the first Polynesian settlers pale alongside the upheaval that has taken place since Cook's arrival. Drive Hawaii's highways today and you will see hardly a plant or tree, let alone a forest, that is a part of the archipelago's historical flora. The most common birds—red-crested cardinals, common mynas, the sweetly singing white-rumped shamas— are overwhelmingly introduced species, and feral cats, pigs, rats, dogs, chickens and mongooses are common even in remote areas. Invasions have supplanted invasions: the Pacific rat that arrived with the first Polynesians is losing ground to the even more voracious European black rat and brown rat, while Polynesian chickens have retreated under the onslaught of an Asian variety. Hawaii once had an incredible diversity of land snails, some 1,500 species like jewels in the forest; there still are land snails on the islands, but they are mainly invasive species, including one, the rosy wolf-snail, that has been killing native land snails since 1955. Every snake, lizard, honey bee and ant you might encounter has been introduced. Hawaii is so different from what it once was that the red hibiscus, an introduced species from China, has become a kind of de facto state flower; the official state flower, the yellow hibiscus or *ma'o hau hele*, which once covered huge areas on every island, is now so highly endangered that it is largely forgotten.

It was amid this tidal wave of change that the Hawai'i mamo finally disappeared. Mamos were seen, and shot, by the Cook expedition's naturalists, who wrote of Hawaii's birds in tender tones—one noted that "the Woods are filled with birds of a most beautiful Plumage & some of a very sweet note," while another declared them "as beautiful as any we have seen during the Voyage." The birds had endured because the Hawaiians managed them with care. Hunters did kill birds for the stew pot—mamo is said to have been delicious—but also used catch-and-release techniques. They applied a smear of latex sap from a native tree that would cause a bird to stick to its perch until the hunters could pluck the colourful feathers, clean the bird's feet and set it free.

Cook's naturalists made note of most of Hawaii's most colourful birds—they were apparently common enough to be readily seen, and the crew also bought many of them, alive, from Hawaiians. More than twenty of these bird species have since gone extinct. The most recent disappearance—a finch-like bird named the po'o-uli, or black-faced honeycreeper—took place as recently as 2004. The causes run the gamut of ecological shocks, among them the spread of avian malaria after mosquitoes were introduced to Hawaii, possibly hatching from drinking water aboard a tall ship in 1826. By the time the last po'o-uli had disappeared, the mamo had been gone for a century or more. The last reported sighting was a small flock spotted in 1899 by one H. W. Henshaw, who shot one of the birds. "It was desperately wounded, and clung for a time to the branch, head downwards," Henshaw would later recall. "Finally, it fell six or eight feet, recovered itself, flew around the other side of the tree, where it was joined by a second bird, perhaps a parent or its mate, and in a moment was lost to view." At the time, Henshaw was working as a specimen collector for the British baron Walter Rothschild; one Hawaiian bird guide-book would later note that the best illustrations of the mamo and several other extinct Hawaiian birds are Rothschild's paintings of his slaughtered subjects.*

It's tempting to picture the Limahuli Valley as a secret paradise, somehow free of the agonies of history. Not so. Kaua'i never had mamos, but the island had its own royal bird, the 'o'o, which had yellow feathers nearly as bright. The last 'o'o was seen in 1985. The woods that climb Limahuli's slopes today are dominated by invasive species; one of the trees they have largely supplanted is the latex tree formerly used by bird catchers. Another plant that has disappeared from much of its former range is *Psydrax odorata*, a shrub with no common name in English. Its Hawaiian name, *alahe'e*, translates roughly as "octopus fragrance," yet the plant's clustered white flowers have a delicious, slice-of-orange scent. The name, it turns out, is an ecological anachronism, a reminder of a time when the flowers grew in such groves that their odour spilled out each morning and evening, slipping in and out of the valleys the way an octopus slips in and out of its hole in a reef. There's no longer any place in Hawaii where the land exhales this particular sweetness. That experience is, for the moment, extinct. The mamo, the 'o'o, the twilight perfume of the alahe'e—double disappearance after double disappearance.

♦ ♦ ♦

Each year, work crews clear another few acres of invasive forest in Limahuli and replant the ground with native species.

---

* Rothschild went on to gather the largest zoological collection ever amassed by an individual, including 300,000 bird skins, 200,000 bird eggs, 2,250,000 butterflies, 30,000 beetles and thousands of skins, bones and taxidermies of mammals, reptiles and fishes. He was famously photographed running a carriage through London pulled by zebras, and also wearing a top hat while astride a Galápagos tortoise.

What they create is not a replica of the original woodland. For one thing, that ancient forest was shaped over centuries by dozens of animals, from tiny snails to flightless geese, that cannot be brought back from extinction without entering the realm of science fiction. For another, there is no hope of rooting out every new plant—more than a thousand species introduced to Hawaii since humans arrived are now free-living in the wild. Still, the dividing line between the invasive and the restored landscapes is stark. The non-native forest looks skeletal, with pale tree trunks rising above a jumble of stones that have lost most of their groundcover plants, and with them, the soil. Where the native woodland begins, the outburst of life is startling, a bright, multi-layered canopy that is green from the treetops to the forest floor. It would be false to describe one as the new Hawaii and one as the old: both are new, though only the restored forest acknowledges the value of the old. The intention in Limahuli is not to live in the past, but to end the war of the present against the past.

Kahimoku Pu'ulei-Chandler, better known as "Mokuboy," is among those who've worked on the restoration crews. He's a young man with the air of a gentle giant, and wears his hair in a wild afro—a style that recalls the bird catchers who once carried out their work in Hawaii's spiritually perilous uplands. Mokuboy, however, is not from the lineage of bird hunters. Instead, his family was entrusted with the sacred fireworks.

Kaua'i was once famous throughout the archipelago for its fire-throwing ceremonies, and the peak of Makana, where the Limahuli Valley meets the sea, is one of only two sites on the island where they were performed. Near the mountain's base was a school dedicated to the perpetuation of the chants, dance and other rites that carried forward history, mythology and genealogy among people who didn't use a written language. When a class of knowledge-keepers was ready to graduate or on other special occasions, fire throwers would climb Makana by night on a secret trail, carrying bundles of dry wood cut to the length of a spear. At the summit, they'd build a fire and chant a call to the wind. At last, the fire throwers would light their wooden spears and hurl them into the air that howled up the face of Makana, curling back on itself to carry the fireworks out over the ocean. The ancient rite, known as *'oahi*, was abandoned in the 1920s; Mokuboy's grandfather had been among the last to carry it out. It was a particularly poignant loss: the forgetting of a tradition dedicated to the celebration of memory.

Mokuboy was my guide into the upper gorge of Limahuli. The valley is his family's traditional ahupua'a; today, both he and his father are employees of the botanical garden. Mokuboy is in many ways a typical young man, spending much of his free time surfing, listening to music and taking photos on his iPad. At the same time, he sees nothing discordant about pausing at the head of the rough trail into what is now a forest preserve to chant an acknowledgement of the simple fact that we, two human beings, are fragile and temporary visitors on a landscape that rose out of the sea five million years ago. "Feels like home, you know?" he said to me afterward. "I feel real comfortable here."

Even before he learned that his ancestors had been fire throwers, Mokuboy had wanted to climb to the peak of Makana. At last he went up the mountain alone, but was forced to turn back after finding himself on too dangerous a slope. From his high point, though, he had seen a different route that could take him to the top. On New Year's Eve, 2011, Mokuboy, his father and a cousin ascended to the summit, this time carrying bundles of sticks that they strapped to their backs with strips of bark. Then they waited for nightfall.

The wood they took up the mountain that night was *hau*, also known as sea hibiscus. It's a magnificent plant—a shrub that can form tangles as big as a house and produces dazzling flowers that are red when they open, yellow as they age and then red again as they fall by the thousands, face up, to decorate the surface of passing streams. Historically, hau was considered good tinder for fireworks, but not the best. The best was *papala*, another plant that has never been given a name in English. Papala wood is hollow and therefore lightweight, making it easy to carry and to catch in the wind, and when one end was lit the flame would pass through the shaft to emerge from the other end as a tail of fire. Today, papala is an endangered species.

Far up the Limahuli Valley, where the waterfall tumbles out of the realm of the gods, Mokuboy pauses beside an unremarkable bit of greenery. As he begins to speak, he holds one of its large, oval leaves; the two of them give the impression of a brother and sister holding hands. The plant is a papala, one of a number planted in Limahuli by the restoration teams. He hopes, he tells me, that papala will one day be abundant enough to be carried again up Makana, if not by him, then by his children or grandchildren.

The first fireworks ceremony on Kaua'i in a hundred years took place unannounced, he tells me. They lit their fire, spoke to the wind and then threw their burning spears into the night. "People were screaming down below, just screaming at the wildness of it," Mokuboy says. Only a lucky few saw it, the flames and sparks riding out into the blackness of sea and sky. It was not a historical reenactment. This was memory come back to life.

Chapter 12.

THE LOST ISLAND

———— · ♦ ♦ · ————

Islands have always been good places to contemplate our relationship to the earth as a whole. Every island is a world unto itself, and the world is ultimately an island. Islands can, as in the case of Hawaii, provide insights that may be useful to the future of the planet. They can also provide our greatest warnings of what may befall us if we fail to change our relationship with nature.

The story of Easter Island is the most celebrated of these cautionary tales. Small enough to get lost in any major city, it is one of the most remote locations in the world, jutting out of the South Pacific some 1,600 kilometres from the next inhabited island and 3,200 kilometres from the nearest continental mainland. Like Hawaii, it was settled first by Polynesian sailors. If the two stories begin the same way, however, they play out much differently.

As the Polynesian population increased, Easter Island's forests—millions of trees—began to decline. Faced with dwindling natural wealth, the people turned to the spirit world for protection. They dedicated themselves to the construction of *moai*, the haunting, hollow-eyed stone heads, the largest standing twelve metres tall and weighing a hundred tonnes, that have become symbols not only of Easter Island but of the pure essence of ancient mystery. Nearly one thousand of the statues were ultimately carved, all of them, incidentally, staring not out to sea but landward, toward the people themselves. Said to have been rolled across the countryside on conveyor belts of logs, the production of moai increased the rate at which the woodlands were cleared. Perhaps the most enduring image from this history is that of the island's last tree. At some point, common sense would tell us, the Easter Islanders must have found themselves with only a single tree left standing. Too far gone to restrain themselves, they chopped it down. Within a generation or two, the idea that there was ever a forest would have sounded fanciful, if not plainly unbelievable. Children would have wondered: Could there ever have been plants that grew up to touch the sky?

There is no happy ending. Struggling to survive on an increasingly barren landscape, the Easter Islanders descended into factionalism, war and human sacrifice. Cannibalism was common; anthropologists have noted that to this day a local insult translates as, "The flesh of your mother sticks between my teeth." When James Cook became the third European captain to stop over on the island, in 1774, his ship's naturalist estimated that it was home to just seven hundred people, all living a desperate existence, their canoes nothing more than patched-together fragments of driftwood. As a society, Easter Island had collapsed.

That's the familiar story.

A competing narrative is emerging, with a more subtle, per-haps even more chilling, message. Recent research suggests that the Polynesians did indeed arrive and clear swaths of trees to make way for their crops, but that they may not have been solely responsible for the demise of the woodlands. The island's trees were probably similar to the Chilean wine palm, which can stand more than thirty metres tall and is the largest palm in the world. Nonetheless, these big trees were vulnerable to seed-eating rodents, which had not lived on the island ahead of human arrival. If the introduced rats were the unstoppable force in the forests' decline, then the island's last tree may simply have grown old and decayed, or blown down in a storm, leaving no seedlings behind.*

The Easter Islanders themselves likely played a leading role in other disappearances. At least twenty major forest plants, six land birds and several seabirds went extinct during the Polynesian era on the island. Yet research now suggests that society did not col-lapse at all. Instead, the islanders adapted and carried on, living in more and more barren surroundings. When archaeologists stud-ied discarded animal bones to determine what the Easter Islanders ate, they found that 60 percent of the bones came from the intro-duced rats. The islanders also developed what is known as "lithic mulching"—rock gardening, essentially. Using lava stones that gradually released their nutrients and sheltered young plants from the harsh elements of the treeless landscape, they managed to

---

* The extinct palms appear to live on in a tree-shaped rune used in the Easter Islanders' Rongorongo script, which was carved into the wood of a shrub imported by early Polynesian settlers. Rongorongo is no longer understood by any living person.

produce enough food to sustain a population density similar to places like Prince Edward Island, Sweden and New Zealand today. Adequately fed, the people used their leisure time to quarry and carve their moai and build grand altars for communal rituals. When the first Europeans reached the island—members of Dutch explorer Jacob Roggeveen's expedition, in 1722—they reported an impressive two to three thousand inhabitants. The people of Easter Island, which is now often referred to by the Polynesian name of Rapa Nui, appear to have been living comfortably, and were more eager to trade or even swindle for hats—no one quite knows why—than for food or any other item. Tests performed on skeletons of Easter Islanders from this era suggest they suffered less malnutrition than people in Europe did at the time.

"Rather than a case of abject failure, Rapa Nui is an unlikely story of success," write Terry Hunt and Carl Lipo, two University of Hawaii anthropologists who've led research on the island since the late 1990s. They point out that the Easter Islanders used human ingenuity and perseverance to build a lasting culture as their surroundings turned empty and desolate.

The question of what really happened on Easter Island is now hotly disputed between collapse believers and resilience theorists, but for our purposes it hardly matters. What the Easter Island stories represent are the two possible end points for our global culture if we continue on our current course toward an ever more simplified and degraded natural world. In the first telling, the fates of nature and humanity are entwined and both go down together in a social and ecological catastrophe. In the second telling, human and non-human life take different paths. The planet's ecosystem is reduced to a ruin, yet its people endure, worshipping their gods and coveting status objects while surviving on some futuristic equivalent of the Easter Islanders' rat meat and rock gardens.

A question lingers: If the decline of the natural world didn't reduce Easter Island to the pitiable state witnessed by Captain Cook and his crew, then what did? The answer is a familiar one. By the time of Cook's arrival, the islanders had been suffering lethal epidemics of European diseases at least since a Spanish ship's visit four years earlier, and possibly since Roggeveen's stopover five decades before. With as many as two out of every three of its people dead, Easter Island's society finally began to fail. Perhaps a lethal pandemic or some plague carried by an alien life form will one day reduce our own global society to chaos in a similar way. No one can say. But if you're waiting for an ecological crisis to persuade human beings to change their troubled relationship with nature, you could be waiting a long, long time.

People still live on Easter Island, including descendants of the original islanders, who never did, after all, go extinct. Their enduring presence is a testament to the human spirit. The natural world of the island, however, has only continued its slide. Birds other than chickens are now essentially absent. The island's flora is heavily grazed by sheep and goats, and introduced weed species outnumber native plants. Visit Easter Island today, and the most visible free-living wild creatures are a few invasive species of fleas and flies.

◆ ◆ ◆

It took centuries to reduce tiny Easter Island to what it is today; it has taken tens of millennia for our species to transform the planet as a whole. Even if we reversed course as of this moment, suddenly striving for a wilder world, it would be years before we could count our first successes. Standing on the globe as we know it today, among people who are predominantly urban, who often spend more time in virtual landscapes than in natural ones, and who in large part have never known—do not have a single personal memory—of anything approaching nature in its full potential, it is hard even to wrap one's head around where to begin.

So: we might begin with an island. Not just any island, but the last major landform on earth to be discovered. It's a rich and fertile place, happens to be large enough to sustain tens of thousands of people, and yet has no history of human occupation. It also doesn't exist, of course, outside of the imagination. Let's call it Lost Island. It looms into view wherever fisheries scientists pass around a bottle of whisky or field biologists stay up late around a campfire, or, for me, every time I stand in a strangely empty landscape, wondering what it looked like when it was full of life.

We've always needed such phantom islands, from the Greek vision of Hyperborea, so remote that disease and old age never reached its shores, to the never-ending search for Atlantis. Buss Island, named for a kind of fishing boat, was described by those who claimed to have explored it as "a champion country," "fruitful" and "full of woods," and appeared on maps of the North Atlantic for nearly three hundred years, despite the fact that all that could be found at its coordinates was deep, cold water and rolling waves. Even today, the world of islands can surprise us: after a close look at the latest satellite imagery in 2011, the number of known islands worldwide increased by 657, while entirely new landforms still emerge from the sea in places like the Gulf of Bothnia, between Sweden and Finland, where the ocean floor is still slowly rising after shedding the weight of its ice-age glaciers. What we seek in such places is the empty slate that allows us to dream of what might have been, or what could be. Or both.

◆ ◆ ◆

Suppose we approach by sea. Perhaps the first sign we are nearing land is an odour, like the rosemary flowers that could once be scented 150 kilometres off the coast of Spain, inspiring the essayist John Evelyn to suggest the plants be grown in London to freshen the air against the smog—and this in 1661. Soon wheeling seabirds can be seen streaking back and forth from the land, and sea turtles—why not?—are making their steady way to their nesting beaches, their shells numerous enough in places that a child could play hopscotch over the waves and never get her socks wet. By the time the dark smudge of landfall finally appears on the horizon, the lines of birds have gathered into clouds, and at times you need to shout to be heard over their cries or take cover below deck from the steady rain of guano.

In places, the ocean seems to hiss and boil as shoals of small fish are pressed to the surface by hungry predators, which fill the deeper water with shadows. "Life water," as fishermen once said. There's no need to bait a hook here: fish will bite at almost any object that appears before their eyes. Navigation, on the other hand, is complicated. The reefs are explosions of colour, as if a crowd had opened a thousand bright umbrellas beneath the sea. Reefs are common around islands, of course, but here they rear up in unfamiliar forms where timeless layers of shellfish have piled one upon the other, rising into ragged ridges. Around and among them bob seals and sea lions; whales blow their rotten-fish breath into the air.

On shore, it's hard to decide whether the island is a thing of rock and dirt or a living being itself. By morning or evening, the birdsong builds into a cacophony, as though a gale was blowing through the wires and bells of a harbour full of wooden ships. Moving inland is surprisingly easy: wildlife trails bore through the forests and traverse the grasslands. The overall impression is not so much of wilderness as otherworldly design. In the same way that modern archaeologists sometimes struggle to differentiate between ancient earthworks made by men versus those built by beavers, our Lost Island has been intimately shaped by its plants and animals in all their abundance. Birds prune the trees as they gather sticks for their nests; moles and boars turn the earth; big, swimming beasts clear routes through weedy swamps.

There are mammoths and sabre-toothed cats, giant camels, giant lizards, giant parrots, giant tortoises. Let us note that on Lost Island you wouldn't feel the freedom that people often do in wild spaces today. You wouldn't swim among the reefs for fear of sharks, and neither would you happily walk alone across the land. You'd quickly learn to listen for a sound in the reeds like fingers gently pulled along a blackboard—a snake—or, worst of all, the sudden hush that falls in the forest when something with fangs and claws is on the move. The ground underfoot would be a busy city of insects, while the air would hum with trilling wings. The mosquitoes—my god, the mosquitoes. If early explorers' accounts from other remote regions of the world are any indication, then Lost Island's biting flies could be bad enough that dogs would bury themselves in the ground and horses run themselves to death, while human visitors would build fires and live in the smoke to escape the swarms.

But the mosquitoes are also food for dragonflies, for tadpoles and frogs, for fish fry, for bats, swifts and swallows. There is, as George Perkins Marsh wrote more than a century ago, "nothing small in nature."

♦ ♦ ♦

From the moment we set foot on Lost Island, it will never be the same again. There's no way to freeze the nature of a place in time, just as there is never a way to turn back the clock to some exact and perfect condition from the past. We can only ask ourselves new questions.

How do we live in a wilder world? And what is the wildest world we can live in?

The first thing we would want to do is establish a baseline. What lives on Lost Island? In what numbers and relationships? What is the total weight of living things? What impression does this wild place leave on our senses, our psyches, our souls? Against the backdrop of deep time, this baseline would be nothing more than a snapshot in a long history of transformation. On human time scales, though, it would be a pole star against which to measure change, recognize losses and make restorations when damage is done. Above all, a baseline is a record of what is possible—a memory, in case one is ever needed in the future, of the abundance and diversity of which nature is capable.

We would want to protect some of that heritage, forever. This is something that we, as a species, have always felt a need to do, whether we've explained it to ourselves through religion, aesthetics, stewardship or a moral responsibility toward other living things. We do this not to divide people from nature but to anchor ourselves in it, to provide sanctuaries where we can witness the natural world without us, as it was during the vast majority of time since the dawn of life on earth. It takes a living planet to sustain us; to ignore the desire to preserve it would be the true beginning of our end.

It also seems to be human nature, however, to kick against the pricks—we don't easily give in to restraints. The *Convention on Biological Diversity*, the most ambitious international conservation accord to date, sets targets that would see 17 percent of the earth's land and 10 percent of the oceans protected from degradation, for a combined total of about 12 percent of the planet. But the agreement goes further, calling on nations to preserve *each type* of ecosystem, from different kinds of forest and grassland to tundra, deserts, mangroves, wetlands, coastlines, even stretches of open ocean. It remains a distant goal.

On Lost Island, we could surely agree to meet these targets: no less than 12 percent of the land and sea, safeguarded for all time. In the past, we would have saved a fragment here and a fragment there, a patchwork, but with today's understanding of the "island effect" we would create the largest, most contiguous protected areas possible, linking them with corridors of wilderness to allow species to move freely across the land and sea.

Then would come the hard part: figuring out how we should live on the other 88 percent.

♦ ♦ ♦

Does it strain your credulity that we would open Lost Island to exploitation? Do you imagine that today's enlightened society would see such an unspoiled place as sacred? Some islands in the Arctic are, in a sense, "lost islands": parts of the Queen Elizabeth archipelago, Canada's northernmost island group, are home to extraordinarily hardy species and remarkable abundance in some seasons, have changed little over the past centuries, and may go decades without witnessing a single human footfall. Today, though, global warming is making the Arctic more accessible, and there's little indication we'll forgo the opportunity. A resource rush is underway, especially for oil and gas. It will be one of history's bitter ironies: catastrophic climate change in one of the world's most changeless places, making it possible to extract more of the fossil fuels that caused the disaster in the first place.

Our Lost Island is irresistibly rich with resources, among them stands of ancient forest and teeming fisheries beyond anything known to most of the world today. But there is also the opportunity to move forward with more conscious aware-ness than we have in the past. Consider the fact that most modern human beings eat next to nothing that is hunted or gathered from the terrestrial surface of the earth. This is an outcome that would strike our ancestors as bizarre if not apocalyptic,* and yet it can't be said to be the product of choice. We drifted to this point, generation by generation.

Not far from where I live today, a seasonal lake—always large but expanding every spring to cover 120 square kilometres—once supplied the local indigenous people with industrial-scale harvests of wapato (WAH-pah-toe), an aquatic tuber that

---

* During early colonial times, some indigenous people reportedly expressed concern about eating pigs or cattle as alternatives to deer, suspecting that people take on the char-acteristics of the animals they eat. Given modern rates of obesity, their fear may have contained a kind of truth.

tastes like a cross between a chestnut and a potato. Wapato were also a favourite food of swans and canvasback ducks, once familiar birds on North American dinner tables. They joined rafts of other wildfowl on the lake, tens of thousands of birds through the seasons, while beneath their dabbling feet throbbed runs of salmon; of oolichan, a smelt so rich in oil that a dried fish can be stood on end and lit like a candle; and of round whitefish, which poured up the feeder streams so thickly they could be caught by hand. The dead from all of these spawning rounds fed sturgeon that could weigh as much as horses.

Despite this wild abundance, the local tribes called the first European settlers *Xweilitum*, "the starving." These were nineteenth-century colonists, often coming from places already deeply disconnected from nature and eager to replace the unfamiliar landscape with fields and farms, with general stores selling flour, salt, sugar and coffee. By the 1920s, some had come to treasure the lake as a larder, but in the end it was drained to make way for agriculture. For years afterward, living sturgeon turned up under the farmers' ploughs in marshy areas of the new fields, spawning fish gathered at the pumphouse that had emptied the lake into a nearby river and flocks of waterfowl landed on the ground as though it still was open water. The indigenous people who relied on the lake continue to feature it on their maps, as though its imprint has survived its material destruction, but the former waters are otherwise forgotten, buried ever more deeply beneath suburbs that act as a capstone against remembrance. Today's residents drive to supermarkets for their food. What was lost with the draining of the lake was not wilderness in the name of civilization, but one more way to feed ourselves.

Would we make the same decisions today? Or, faced with the knowledge of what nature can be, would we strike a different balance between whole ecosystems that feed every living thing and simplified landscapes that feed nothing but ourselves? Today, the only wild-caught foods that most people eat are fish and shellfish. How many of them would we take from Lost Island? "To have a reasonable chance to feed people with seafood over the next fifty years, we need to understand these things," says Tony Pitcher of the University of British Columbia Fisheries Centre. No one yet has developed a model of the historical abundance of all the world's oceans, so it's impossible to say how many fish we could have caught from those pristine seas without overfishing. But Pitcher, whose research has taken him around the globe, says his back-of-the-napkin estimate is that, if we miraculously rebuilt the planet's fish stocks to their original bounty, we could catch 10 percent of the fish every year without any overall decline in abundance. That amount of fish, he guesses, would equal between 40 and 60 percent of the annual catch worldwide today. It would surely be a shock to lose half of today's fishery. It would take innovation and adaptation to close the gap in the food supply—but nowhere near as much effort as it will take to face the future as our current fishing practices deplete the last of the world's major fish stocks. The upside, meanwhile, would be a supply of seafood that never decreased, while our coasts and oceans exploded with life.

As we try to build ourselves into the nature of Lost Island, raising cities, towns, farms, mines, ports—all the trappings of our lives—these are the kinds of questions we must grapple with. And the solutions will not be familiar ones. Maybe on Lost Island there are still herds of wild bison, and we eat as much or more of their meat than beef. Maybe we harvest both wheat and wildflower bulbs. Maybe we acquire our former taste for porpoise, seal and whale. Maybe we get to know the flavours of songbirds, and in exchange spare more forests and grasslands from our bulldozers and ploughs.

♦ ♦ ♦

If the goal on Lost Island is to live with all the wild diversity and abundance we can, then some acts that are common today would not have any place there. Large-scale clear-cut logging of ancient forests, which reduces woodlands thousands of years in the making to barren fields of stumps, would not be practised on Lost Island, and neither would its marine equivalent of dragging weighted trawl nets across the sea floor. Other practices would be carried out far more carefully than in the past: fully a third of the world's largest river systems, for example, have been strongly affected by dams, cutting off major fish migrations between the fresh water and sea. We wouldn't use waterways as dumping grounds for toxic waste or raw sewage, build cities at river mouths, which are some of the richest ecosystems on earth, or drain swamps because we consider them a nuisance. Mountaintop removal and other forms of mining that tear away huge areas of the earth's surface rather than tunnelling underground would be practised in only the most extraordinary circumstances, if at all. In areas of especially high biodiversity, we might even decide to leave billions of dollars of fossil fuels in the ground, as the nation of Ecuador is planning to do—at significant economic sacrifice—in part of the Amazon Basin. Replacing whole horizons of grassland or forest with the unbroken fields of industrial farming would be unthinkable. The idea of displacing seabird colonies, sea turtle nesting sites, fish spawning beaches and seal haulouts simply to provide vacation homes for the rich would beggar belief.

These are the low-hanging fruit: the human activities where the highest environmental impacts have been embraced unnecessarily. In every case, alternatives already exist to meet similar needs while causing far less damage, typically by employing more people, using materials more efficiently, or applying more sophisticated technologies. We are talking, however, about acts of self-restraint already resisted by the world's most powerful corporations and their many stakeholders, and the questions only become more difficult, if not overwhelming. How do we handle plastic on Lost Island, knowing it has been found in the digestive tracts of nearly three hundred different marine species and that toxic chemicals from degrading plastic turn up in samples of ocean water all over the world? Do we build wildlife overpasses and underpasses on every stretch of highway, and do we build fewer roads overall? Do we live in darker cities, knowing that our bright lights threaten migratory birds, cause some flowers not to bloom and some animals to mate out of season, lead bats to starvation and leave fireflies unable to signal potential mates out of the night? Can we accept that a recreational hike in the Lost Island wilderness may demand a minimum of five people, possibly armed? Are we able to think beyond our own senses, recognizing that most birds, for example, can see light in the ultraviolet spectrum, that pigeons can hear the infrasound rumble that precedes an earthquake, that bears can pick up odours the way we can hear a pin drop in an empty room? Will we eliminate all those mosquitoes?

Life on Lost Island would quickly convince us that we truly cannot live in the past; we always and only exist in the present. Perhaps, in our largest, most pristine reserves, megafauna still would march behind perimeter fencing, but most of the landscape would be more nuanced and imperfect—more human—than that. For most of our history as a species we have been running a largely unmonitored, planet-wide experiment in reinventing nature; still, the way forward is not to bring an end to experimentation, but to proceed more carefully and consciously. We cannot choose not to choose.

There are ordinary living things among us—cockroaches, crocodiles, the simple plant known as horsetail—that have endured, nearly unchanged, since the age of the dinosaurs; horseshoe crabs, which look like leather luggage from another planet, have been around several thousand times as long as human beings and have survived five mass extinctions. More important than any one species, however, is the system that created us, nature, with its roots that reach back more than three billion years. Even in the heart of a city of millions, even where we dig a mine or position a deep-ocean drilling rig, we can do more to accommodate the living earth. It is inspiring—perhaps even liberating—to acknowledge that we have the power and the necessity to shape the future. But we have been attempting to make an impossible world, in which humans are separable from the rest of life. Our greatest experiment is still pending: the making of a world in which humanity can express all of its genius, and so, too, can nature. Our Lost Island is not life as it was before the Industrial Revolution, or before Columbus, or before humans walked the earth, but a way of being that has yet to be invented: a world true to the past and unlike anything seen before.

◆ ◆ ◆

A few words about hubris:

In around 1820, European sailors introduced cats to Macquarie Island, which hangs like a frozen tear on the face of the Southern Ocean between New Zealand and Antarctica. Treeless, cool and wet, the island nonetheless was home to unique birds, one of them a brightly coloured and seemingly misplaced parrot called the Macquarie Island parakeet. Already you can guess how this is going to end: the cats drove the flightless parakeet extinct. But there is more to the story than that.

As the number of cats slowly grew, the parakeets lived on, nesting under tuffets of greenery and feeding in running flocks along the shoreline. In 1872, men who camped on the island to kill penguins and seals introduced a game bird from New Zealand called the weka. Wekas will eat almost anything, including parakeets, but the little birds continued to endure. They were clever survivors, and had the seasons on their side. Every winter Macquarie's millions of seabirds and penguins would migrate away, leaving the cats and wekas hungry. There was never a steady enough food supply for the predator population to endlessly grow.

Then, in 1878, sealing crews brought rabbits to the island. As the rabbits multiplied, there was suddenly plenty of food for the cats and wekas all year long. Their numbers surged, and the parakeets went extinct within a decade.

Flash forward one hundred years, and Macquarie Island was wearing a living, breathing rabbit-fur coat, with a thousand rabbits per square kilometre. Their intensive grazing threatened to wipe out every trace of the island's original vegetation, so wildlife managers introduced yet another new species: fleas that spread a virus lethal to rabbits. As the rabbit population decreased, the native plants began to recover. Then biologists noticed that Macquarie's seabirds were dying off. The culprit was the hungry cats and wekas, which had become reliant on the rabbits as food. A program was launched to eradicate the predators, with the last weka killed by 1989 and the last cat shot in the year 2000. But as the cats and wekas disappeared, the rabbits boomed again: it turned out that the disease-carrying fleas were not enough to control the population once the predators had vanished. Soon there were so many rabbits that not only were they once again destroying the plant life, they were also invading the nesting sites of the seabirds that had been rebounding after the elimination of the cats.

Macquarie Island is only thirty-five kilometres long, just a rock in the sea, and yet we failed again and again to foresee the consequences of our actions there. R. H. Taylor, the ecologist who first solved the riddle of why the parakeets had endured the predatory cats and wekas only to vanish with the introduction of vegetarian rabbits, came up with a lone recommendation that remains, really, the guiding sutra for the new wilderness of the future: "We should watch for new factors."

Who will be on watch? The people who would live on Lost Island are not figures from the past—they are us. In some cases, the losses we have suffered from the decline of the natural world are still fresh. The nation of Malawi in southern Africa, for example, frequently appears on top 10 lists of the world's poorest nations; malnourishment is widespread, and people have been imprisoned for such crimes as stealing leftovers from their employers' dinner tables. As recently as the 1980s, however, the fishery for a wild whitefish called chambo

in Lake Malawi, one of the largest water bodies in Africa, supplied twice as much protein per capita as Malawians eat from all
animal sources today. After decades of overfishing, the chambo
stock collapsed in 1991 and has not recovered. Malawi is a
nation that is literally hungry for nature as it was.

Yet for many of the most privileged people on earth, and I
count myself among them, whatever ecological wounds we may
have suffered have faded into scars. To fly with ease between the
world's great cities; to hurtle through rushing streets among
people from every corner of the globe, music in our ears, emo-
tions modulated by pharmaceutical drugs; to communicate
across the planet with the touch of a finger on a screen, can feel
as mystic, as wild, even as sacred as all but the most extraordinary encounters with the natural world today. The global majority who live in cities, whose families may have been urban for
generations now, are often transient, strangers to their landscapes, temporary visitors with no place that is truly home and
no traditions in the places they find themselves.

The history of nature tells us we have been a part of a great
forgetting, and can now be a part of the reminding. What is the
twenty-first century equivalent of the child who can swim with
sharks, or the gatherer whose understanding of plant biology
is the equivalent of a lifelong scientist's, or the artist who can
capture uncertainty in the face of a bison with a few deft lines
of fingerpaint, or even the working sailor who knows the
unmistakable joy that can be expressed by a tortoise? These
questions await our reply. To live in a wilder world, we'll have
to find a way to weave nature into our identities, until guarding
against harms to the natural world is as innate as watching out
for ourselves, our families or our communities. Only this kind

of person—we might call him or her the ecological human—can inhabit nature deeply enough to change our troubled relationship to non-human life, to observe carefully enough the changes we will continue to make, and to truly love the return of the wild as a formidable presence in our lives.

For some, such a transformation is probably impossible; they have been away from nature too long and in too many ways to make their way back. But the world has never needed every soul to act as its guardian; it has only ever needed enough. Most of us—a great majority, I hope—can still take some delight in the awkward movements of a newborn owl or the intricate architecture of a flower. It is this capacity that now matters most to our future as a species: the part of us that feels awe in the knowledge that a simple clam, *Arctica islandica*, can live for as long as four hundred years, that the gingko tree has remained essentially unchanged through 280 million years of evolution, but also that some insects have adult lives so brief they are born without mouths to eat with. There is a person within us who would like to hear birdsong spill out of the forest like a wave, watch spawning fish turn a blackwater river to silver, or walk a road beaten into the savannah by herd animals. It's that same person who would take some unexplainable satisfaction from the sound of a whale's deep breathing as it sleeps at the surface of the sea, and who is able to grasp that a lichen that clings to the slopes of a single mountain is a metaphor for our own dependence on this lone earth in outer space.

What we do not know—what we have not even begun to ask—is how many people can live on a rewilded planet. Here, though, is an interesting coincidence. The highest credible estimate for the population density of Hawaii during the era in which the islands' people survived in isolation from the rest of the world is forty-eight people per square kilometre. Compare that to the planet's current population of seven billion people, or about forty-seven people per square kilometre of land. At the risk of pointing out the obvious: the two figures are almost exactly the same.

Most of the earth is nothing like warm, perpetually fertile Hawaii. Then again, the Native Hawaiians of that earlier time had no metals, no fossil-fuel energy, no trade with other nations, no way to bring home fish from faraway seas—they had lived not one of the lessons or insights of the past two hundred years. They lacked the collective intelligence and technologies of the globe's seven thousand cultures, not to mention supercomputers capable of performing nearly twenty quadrillion calculations per second, and they had no one who could build on their successes or support them through their failures. In other words, it just might be possible for seven billion people—maybe even more—to survive on this planet, and not only to stop the endless decline of the natural world but watch it return to astounding, perpetual life. All it will take is a wilder way of being human.

So move your feet
from hot pavement and into
the grass.
ARCADE FIRE

Epilogue

———— ◆ ◆ ◆ ————

In the history of the grasslands I grew up on, one creature stands out in my mind above all others: the grizzly bear. There lurks the great beast in pioneers' memoirs, in one case sending a prospector's pick "skyrocketing upward" with the swat of a paw, in another killing a man in the pine forest and "eating half his horse." Shamans of the local Secwepemc tribe wore grizzly claws around their necks, and during fifteen years in the mid-1800s, the trading fort that became my hometown received 104 of the bears' pelts. The grizzly is a memorable animal, equally able to bite through a six-inch pine tree or nimbly open a screw-top jar, to run a marathon distance in a single hour or lift and carry 150 kilograms of weight in its mouth, but as a child I never once heard the prairie remembered as grizzly country.

Eliminated from 98 percent of its territory in the Lower Forty-eight and at least a quarter of its former Canadian range, the grizzly is now known mainly from cold mountain meadows and distant northern rivers. I have seen the bears in the wild myself—in cold mountain meadows and along distant northern rivers. I failed to picture them picking cactus stickers out of their paws.

I called Jeff Gailus, the Missoula, Montana–based author of *The Grizzly Manifesto* and a man who can be fairly described as a grizzly historian. He put me straight in plain language: "The grizzly bear is a grasslands animal." The first undisputed grizzly sighting by a European explorer, a man named Henry Kelsey, took place in 1691 near the location of modern-day Preeceville, Saskatchewan—a thousand kilometres east of the nearest grizzly range today, and deep in the heart of Canada's breadbasket. Lewis and Clark saw their first grizzly tracks in 1804 in the Great Plains state of South Dakota. In their first close encounter with an adult grizzly, Clark and another member of the party shot the animal ten times, then watched as it swam half the Missouri River to a sandbar, where another twenty minutes passed before the bear was dead. "These bear being so hard to die rather intimidates us all," Lewis wrote in his journal. The explorers were in a part of Montana now best known for its wheat fields.

Gailus's own research into grizzly history began when he saw a rogue bear loping through the prairie of southwestern Alberta, a visual shock he compares to "a trick of the sun or a fatigue-induced dream." To the few who keep its memory, the grasslands bear is known as the plains grizzly, though it was not a separate species. A plains grizzly was simply a grizzly bear, *Ursus arctos horribilis*. In fact, some of the bears in the Rocky Mountains today are thought to be descended from grassland bears that were pushed westward as the bison were wiped out and the land ploughed under. Some of the bears' open-country traits still linger: faced with a confrontation, they don't look for a tree to climb. They take what's coming, head on. Gailus watched his plains grizzly run straight through a barbed wire fence.

I asked Gailus if there was anywhere on earth where I could still see a grizzly roaming a landscape like the one that I grew up on—where I could witness in reality what my mind's eye could not imagine. I had the feeling I was asking the impossible.

"Go to Yellowstone," Gailus said.

It took sixteen hours to drive across the American West, and I looked out on the passing forests and plains with thoughts of the past. The only large wild animals I saw were birds: a few hawks and herons, a lone bald eagle. The land seemed eerily empty at a continental scale.

That impression changed when I entered Yellowstone National Park through its northern gateway in Gardiner, Montana. Over the next half-hour I saw bison, elk and prong-horn, as well as a mule deer and a lone black bear. Gailus had told me: to see big carnivores, look for people. Whenever a bear, wolf or cougar is spotted in Yellowstone, the roadside quickly turns into a parking lot, complete with uniformed traffic rangers. I found one of these gatherings soon enough, the crowd ogling the liminal distance through telescopes and zoom lenses. The scene was faintly absurd—a paparazzi of the predators—and yet there was something heartwarming, too, in this hunger for a glimpse of a faraway wolf, asleep at the base of a tree.

It was dusk when I moved on, intending to camp in the Lamar Valley in the park's northeastern corner, where I would go searching for grizzly bears the next day at first light. The Lamar is a sagebrush bottomland surrounded by slow-rising peaks—a near replica of my childhood home. Or so Gailus had told me, anyway: my first view was of darkness, my headlights flashing across signs that informed me every campground was full. I found a vacant site, though, on the banks of Slough Creek. A young couple had reserved it, but was hastily packing to leave when I pulled up. I was welcome to take their place, they said. They'd seen the first grizzly of their lives that day, and no longer wanted to sleep in bear country.

◆ ◆ ◆

I know the fear that bears can inspire. I inherited it from my mother, who, when she was a girl, walked to school every day through a windrow of aspen and maples. Every time she did it, she was afraid that a bear would be lurking in the trees.

There was almost literally nothing to fear. My mother grew up where the great central grasslands of North America once began to give way to northern forests, but by the time she was born, there were no real grasslands, only farmers' fields, and no forests, only occasional groves of trees. Outside of the barnyard, there weren't a lot of animals. She remembers deer, the odd coyote and prairie dogs with a bounty on their tails. Once, word spread from farm to farm that a mountain lion had been seen. She can't recall if someone finally killed it, but she can still picture the awestruck men going out to hunt the animal. She prefers to believe that it got away.

My mother grew up among Finnish immigrants, the descendants of people who practised *arctolatry*, or bear worship, whose *Kalevala* creation story speaks of the "sacred bear with honey fingers," and who saw bears as forefathers and foremothers. Given this heritage, my mother might reasonably have considered bears to be something like family, but she never had the chance. On the grasslands as she knew them, the plains grizzly was not only gone but utterly forgotten, and black bears were close to the brink of extirpation. My mother's fear of bears was like a fear of ghosts—grounded not in truths but legends, not reality but vapours. Then, one day, her years of unrequited fright were miraculously rewarded: a black bear was actually spotted in the area. She went out with her family to see it, an animal so black it was as though the beast and its shadow were one and the same. She watched the bear walk the grid roads between the fields, and then disappear—into a windrow.

My mother grew up, married my father, moved farther west, and never allowed her fear of bears to limit her. Still, she passed it down to her children in situations like this one: She was leading me, my brother and a group of our friends on a mountain hike when the chronic pain in her knees forced her, as it often did, to walk backward the final few miles. Moving forward while looking backward is not a comforting way to travel in bear country, so my mother had all of us singing in order to scare any animals away. Without ever saying so in words, she impressed on us that any run-in with a bear would end with us in its stomach. And there we were: a troop of boys belting out the "Heigh-Ho" chorus from *Snow White and the Seven Dwarfs*, led by a pretty, black-haired woman walking backward with bells jangling on her rucksack.

My own fears have become inseparable from fascination. I had my first face-to-face encounter with a bear while I was still a child, riding my bike in loops around a mountain campground. I fled, but so did the bear, and my perspective on the animals began to shift. I have now seen bears in the wild more times than I can count, the vast majority of them black bears, *Ursus americanus*. They are individualistic animals, each one with its own personality, and can run at fifty kilometres per hour, climb trees like a cat, and are so strong that even yearlings have been observed flipping 150-kilogram rocks, backhanded, with a single paw, in order to get at the insects underneath. Their most fantastic achievements, however, may be in the field of play: bears have been witnessed turning somersaults, diving from high places, swimming through hollow logs, doing headstands, throwing rocks, shadowboxing, hanging upside down, swinging from vines, making snowballs and sometimes just quietly sitting and watching other animals.

Black bears have also killed an average of one person every twenty months in North America since the year 1900, making them statistically the most dangerous big animal on the continent.* Everything about large, meat-eating creatures is a challenge to our modern way of life. Most need a lot of land to find food. An average female grizzly in Yellowstone roams a home range of nearly nine hundred square kilometres; males may cover 3,700 square kilometres or more. Polar bears in the Canadian Arctic have been known to wander across some 125,000 square kilometres in a year, an area almost fifteen times the size of Yellowstone National Park. Predators are also often quick to disappear as humans intrude. A world map of the remaining places where brown bears—the species of which grizzly bears are a part—can still be found amounts to a map of the northern hemisphere's most remote and undeveloped regions. Even the American black bear, which lives alongside people willingly enough to be a garbage-eating, orchard-raiding

* For comparison, people in Canada and the United States are 60,000 times more likely to die in an automobile accident.

nuisance in some places, tends to fade as the human footprint increases. Other predators, such as the wolverine, may begin to abandon large areas with the construction of even a single road.

The presence of fearsome beasts is also affected, of course, by the fact that they can threaten our lives. While psychologists have shown conclusively that people from across the world's cultures tend to prefer natural settings over the built environment, that same research does make certain distinctions. As much as we favour savannah-like landscapes, we're also generally wary of dark, dense underbrush or tightly spaced trees. Our modern culture may have inherited a bond with nature from our ancient ancestors, but we prefer a version of nature in which hungry animals with large teeth have nowhere to hide. On that count, we have had our wish. As our increasingly urban and technological lives widen the gap between us and nature, the globe has come to be dominated by a single way of relating to dangerous animals, and that is to banish them to places where few if any people live.

We have always had other options. Around the world, over long stretches of time, communities of people have developed ways of living in close contact with bears, wolves, tigers, lions and leopards—even giant, man-eating lizards and sharks. In the Malay Archipelago of Southeast Asia it was once a widespread custom to encourage the presence of the local *macan bumi*, the "village tiger," with regular offerings of food. The relationship was a complicated one, and probably can't be fully understood at this distance in time, but among the reported results was that even children could drive off a tiger if it strayed too close to a herd of cattle. Anthropologists among the Khoisan people of Botswana, in southern Africa, observed small groups of hunters effortlessly driving prides of lions off their kills in order to take a share of the meat; the lions could easily have savaged the lightly armed men, but did not do so. Similarly, early European settlers in eastern Canada reported indigenous villages that looked on mountain lions as providers—the people would scavenge portions of their kills—and resented it when newcomers shot the big cats. In Hawaii, explorers watched the islands' inhabitants swim out to their ships while huge sharks passed beneath them; at times, a shadow would rise from the deeps toward a swimmer, who would reach down to strike the deadly animal on the snout, sending it on its way. Predators persisted for millennia in places where people were easily widespread and technologically advanced enough to erase them. Lions could still be found across northern India until the colonial era of the nineteenth century, when they faded to a single small enclave, and there were thousands of tigers in southern China as late as the 1950s. In Turkey, one of the oldest civilizations on earth, the last Caspian tiger was shot only in 1970. No one can say exactly what kind of truces permitted the presence of people and predators in so many places, only that today we have forgotten their terms. "Each generation has to renegotiate its relationship to these animals," one bear conservationist told me. "We have to find new levels of meaning in these animals in a world that is increasingly confusing."

Even the disappearance of "man-eating" beasts around the world contains a surprising fact: it isn't fear for our lives that has led to their destruction. Most have dwindled inch by inch, not with a bang, but a whimper—an accumulating loss as we invade and destroy the animals' habitat, slaughter their prey, sell their skins or gallbladders as trophies or superstitious remedies. Often, as with the bounty-hunting manias of the nineteenth and twentieth centuries, the slaughter is aimed primarily at protecting livestock. Only when a predator becomes rare do we tend to root it out as though the beast was evil incarnate. By then, like the bears in my mother's windrow or my own childhood daydreams, what we are truly afraid of is the unknown. It isn't fear that drives us to extinguish fearsome beasts, but once they are gone, it's fear that keeps us from bringing them back.

♦ ♦ ♦

Jeff Gailus had sent me to the right place. As dawn broke in the Lamar Valley, the day's first light played across tussocks of bunchgrass, and the scent of rabbitbrush was rising with the dew. The song of the meadowlark, anthem of the grasslands, rang out above the balsamroot flowers; my brothers and I used to pick them for Mother's Day bouquets. The valley was the living image of the land I had walked as a child.

I drove from my campsite onto the main road, a skiff of fresh snow sizzling under my tires. It was too early for most of the paparazzi, and I had to scan the landscape with my own myopic eyes and inadequate binoculars. At last I pulled over beside a lone truck whose driver was aiming an enormous spotting scope at a dark shape on a nearby hillside. "Black bear," the woman said. "A big one." I stepped up to her eyepiece and squinted. The bear was buckwheat-honey brown, nearly black, but it was not a black bear. The humped shoulder and flat forehead gave it away as a grizzly.

This was it, then: a grizzly bear in the grasslands. In my mind, I had already decided how the experience would play out. I would see the big bruin shouldering through the sagebrush, majestic in the way that only a grizzly bear can be, and I would know in that instant the thrill and awe that could have been a part of my childhood. It would be, I thought, a kind of home-coming—a way of making my own past more true and more whole. Angels would sound their heavenly trumpets, and glorious rays would beam down from the clouds.

Instead, I saw that the bear's jaws were dripping with blood. Every few moments, it would dip its head into the entrails of a calf elk it had killed, while a cow elk, presumably the dead calf's mother, stood a terrible vigil only a few paces away. The bear tugged and tore at the meat, the mother elk stepped in and out, in and out, of the monocular's field of vision, and there was nothing at all familiar about the cold violence and danger that the scene implied. Of course there wasn't: springtime in Yellowstone is not the season of gambolling fox kits, of care-free Mother's Day bouquets, but of the hungry bear and wary bison. Of death, that ordinary horror.

I stood there long and long, as the poets say. It was a spectacular morning. Along a hill's crest, a herd of elk waited for the dead calf's mother to find whatever resolution would allow her to join them and move on. Young bison butted heads on the valley floor, while pronghorn chased each other so effortlessly and tirelessly across the sepia landscape that it was like watching a silent film on an endless loop. Later, after the grizzly had rolled onto its back to digest its meal, revealing hind feet that, at a distance, resembled a human baby's, a pack of wolves came nosing out of the upland pines. The lead wolf was cream-coloured, almost golden under the rising morning sun. This was my home as it might have been. It was magnificent, it stirred the soul—and I knew I was a stranger here. The absence I'd grown up with had not only been around me, but within me as well.

As a boy, I sometimes sat down from my wandering only to wake up an hour later, surprised to find I had fallen asleep in a warm patch of grass. That wouldn't happen in bear country. When I walk in a place like Yellowstone, it's always with a slight but solemn recognition of the slender possibility that I will die, that some wild animal will kill me. My senses come alive: I taste the air, listen for sounds beneath the wind. Suddenly, nature is not the backdrop to life, it is life itself, and I am no longer myself, but myself in nature. I note and classify even small changes: a shrew darting across the path, an updraft twisting a fern frond, a hummingbird gathering spiderweb for its nest. Light and form take on greater clarity, and given enough time to sink into these sensations, visual tricks will arise that are somewhere between vigilance and hallucination, such as seeing clearly every trembling leaf on a tree while in the same moment watching a bumblebee pass by in slow motion. As my senses reach outward, I spread away from myself. The world expands. It's the closest a person can feel, I think, to being a flock of birds.

The naturalist John Livingston described this perspective as a *participatory* state of mind, and speculated that among wild animals it is the ordinary form of consciousness. It would seem to have to be. It's possible, of course, to stumble through the wilderness while locked inside yourself, mentally racing over day-to-day worries, but that is not a good way to remain alive. It's not that self-awareness is absent in animals—it has been tentatively revealed in experiments involving such species as apes, dolphins, magpies, even octopuses—but that it is a less useful tool than an outward mind: to endure among other species, you must experience the world as a place you share with them. "Awareness of the whole self is emotional, not rational," wrote Livingston. "It is lived, not abstracted. It is received, not perceived. It is a gift, not an accomplishment." Twenty years ago, researchers in Banff National Park began using remote cameras to monitor trees that bears were known to rub against. They were surprised to discover that not only bears were visiting these trees, but so was almost every other mammal in the forest: deer, moose, elk, bighorn, mountain goats, wolves, coyotes, foxes, lynxes, cougars, wolverines, martens, squirrels, wood rats, porcupines, even people's dogs. The trees were everywhere—at least forty-two of them were identified along a single, twenty-four-kilometre hiking trail. While the messages that the animals were leaving behind in scratch marks and musk, in hair oils and urine, were untranslatable to our tongues, they were obviously important. Wildness has its own understandings.

Most of us rarely—if ever—inhabit the world in this way. The child raised among foxes is different than the child raised among grizzly bears. I hesitate to say that one is better than the other, but the two unquestionably stand apart. Who might I have been as a son of grizzly country? More anxious in life, or less? Better or worse able to negotiate the price of a used car? More or less likely to believe in the Bible as the literal word of God? I could no more answer those questions than I could say who I would be if I'd been born to different parents. The absence of the grizzly, the presence of the fox—the outcome was decided over so much time that no one even noticed they were choosing.

Once, though, I spoke to a mother and a daughter who had lived almost thirty years in grizzly country. The daughter had been raised there, a twelve-hour drive from the nearest major city, and I asked her what it was like to grow up where grizzly bears roam. It was like asking a child of the city what it was like to grow up among all those cars. She couldn't make sense of the question—couldn't parse into threads the whole she had always known. It was the pattern of ecological amnesia thrown into reverse: we are always only a single generation away from a new sense of what is normal.

Then I spoke to that woman's mother. Her name was Sally, and she had not grown up with grizzly bears. She was fully an adult when she and her husband left New Mexico with that common but usually fleeting longing to live somewhere truly wild. Their search took them across America and finally into Canada, and even then they kept moving north. At last they discovered the Tatlayoko Valley on the remote Chilcotin Plateau of British Columbia, with the glacier-smoothed hummocks of the Potato Range on one side, ragged peaks on the other, and the cool eye of a frigid lake in between. Tatlayoko sits exactly at the line where coastal weather fades into the lee side of the mountains, and it does not always agree to go quietly. There are clear-sky days when wind will rage down onto the water, shatter the surface into droplets, and carry them up the valley as a sheet of instant rain.

Sally had told me a story that I only fully came to appreciate as I stood on the Yellowstone grasslands. It isn't easy to adapt to a life in the wild, she said, especially as an adult. You have to learn your way in. Twice she had close calls. The first time, she startled a grizzly and, reacting without thought, she turned to run. Wrong move. It chased after her, a nightmare that ended only when her little dog, adopted from the streets of the nearest town, burst out of the brush to bite the bear on the nose. The second time, she surprised a sow grizzly and its cubs. Again the bear charged, but this time Sally stood her ground. The animal stopped only inches away, the cubs bawling around its feet. When it roared, Sally could feel the vibration pass through her stomach, through her bones. Then it swiped once with a paw, slicing through two layers of clothing and the skin of Sally's thumb. With that, the mother bear's fury drained away. The grizzly ambled off; the scales of life and death tilted back into balance; the crawl of time returned to its regularly scheduled programming.

"It was really a highly spiritual experience for me," Sally said. She shared this revelation cautiously, aware that it would be hard to understand. But in those terrible instants, she said, she realized that the bear was only doing what it must, and so was she, and so, too, were even the meadow grasses and the trees, the earth and the sky, all of it blurred into a pattern too infinite and ancient to explain. At last, Sally found the words for the feeling: "It was just like coming home."

Selected Bibliography

Taken together, the following sources can be considered this book's compass. For complete notes and citations, visit jbmackinnon.com.

1. ILLUSIONS OF NATURE

Kamler, Jan F., and Warren B. Ballard. "A Review of Native and Nonnative Red Foxes in North America." *Wildlife Society Bulletin* 30, no. 2 (Summer 2002): 370–79.

Lord, John Keast. *The Naturalist in Vancouver Island and British Columbia*. London: R. Bentley, 1866.

Lowenthal, David. *George Perkins Marsh, Prophet of Conservation*. Seattle: University of Washington Press, 2000.

Marsh, George P. *Man and Nature; or Physical Geography as Modified by Human Action*. New York: Charles Scribner, 1865.

Ponting, Clive. *A New Green History of the World: The Environment and the Collapse of Great Civilizations*. New York: Penguin Books, 2007.

Worster, Donald, ed. *The Ends of the Earth: Perspectives on Modern Environmental History*. Cambridge, UK: Cambridge University Press, 1988.

2. KNOWLEDGE EXTINCTION

Assmann, A. "History, Memory, and the Genre of Testimony." *Poetics Today* 27, no. 2 (2006): 261–73.

Braund, Kathryn E. H. *Deerskins and Duffels: The Creek Indian Trade with Anglo-America, 1685–1815*. 2nd ed. University of Nebraska Press, 2008.

Chabris, Christopher F., and Daniel J. Simons. *The Invisible Gorilla: And Other Ways Our Intuitions Deceive Us*. New York: Crown, 2010.

Cohen, Stanley. *States of Denial: Knowing about Atrocities and Suffering*. Cambridge, UK: Polity, 2001.

Fuller, Errol. *Dodo: From Extinction to Icon*. London: Collins, 2002.

Kahn, Peter H., and Batya Friedman. "Environmental Views and Values of Children in an Inner-City Black Community." *Child Development* 66, no. 5 (October 1995): 1403–17.

McClenachan, Loren. "Documenting Loss of Large Trophy Fish from the Florida Keys with Historical Photographs." *Conservation Biology* 23, no. 3 (June 2009): 636–43.

Paddle, Robert. *The Last Tasmanian Tiger: The History and Extinction of the Thylacine*. Cambridge: Cambridge University Press, 2000.

Papworth, S. K., J. Rist, L. Coad, and E. J. Milner-Gulland. "Evidence for Shifting Baseline Syndrome in Conservation." *Conservation Letters*, April 2009.

Pauly, Daniel. "Anecdotes and the Shifting Baseline Syndrome of Fisheries." *Trends in Ecology & Evolution* 10, no. 10 (October 10, 1995): 430.

Ricoeur, Paul. *Memory, History, Forgetting*. Chicago: University of Chicago Press, 2004.

Saenz-Arroyo, A., C. Roberts, J. Torre, M. Carino-Olvera, and R. Enriquez-Andrade. "Rapidly Shifting Environmental Baselines among Fishers of the Gulf of California." *Proceedings of the Royal Society B: Biological Sciences* 272, no. 1575 (September 22, 2005): 1957–62.

3. A TEN PERCENT WORLD

Carlton, James T. "Apostrophe to the Ocean." *Conservation Biology* 12, no. 6 (December 1998): 1165–67.

Ceballos, Gerardo, and Paul R. Ehrlich. "Mammal Population Losses and the Extinction Crisis." *Science* 296 (May 3, 2002): 904–7.

Jackson, Jeremy B. C. "What Was Natural in the Coastal Oceans?" *PNAS* 98, no. 10 (May 8, 2001): 5411–18.

Lotze, Heike K., and Boris Worm. "Historical Baselines for Large Marine Animals." *Trends in Ecology & Evolution* 24, no. 5 (May 2009): 254–62.

McClenachan, Loren, Jeremy B. C. Jackson, and Marah J. H. Newman. "Conservation Implications of Historic Sea Turtle Nesting Beach Loss." *Frontiers in Ecology and the Environment* 4, no. 6 (August 2006): 290–96.

Morrison, John C., Wes Sechrest, Eric Dinerstein, David S. Wilcove, and John F. Lamoreux. "Persistence of Large Mammal Faunas as Indicators of Global Human Impacts." *Journal of Mammalogy* 88, no. 6 (December 2007): 1363–80.

Mowat, Farley. *Sea of Slaughter.* Boston: Atlantic Monthly Press, 1984.

Myers, Norman. *The Sinking Ark: A New Look at the Problem of Disappearing Species.* Oxford: Pergamon Press, 1979.

Myers, Ransom A., and Boris Worm. "Rapid Worldwide Depletion of Predatory Fish Communities." *Nature* 423, no. 6937 (2003): 280–83.

Roberts, Callum. *The Unnatural History of the Sea.* Washington, DC: Island Press/Shearwater Books, 2007.

Roman, J. "Whales Before Whaling in the North Atlantic." *Science* 301, no. 5632 (July 25, 2003): 508–10.

4. THE OPPOSITE OF APOCALYPSE

Barnosky, A. D. "Assessing the Causes of Late Pleistocene Extinctions on the Continents." *Science* 306, no. 5693 (October 1, 2004): 70–75.

Bucher, Enrique H. "The Causes of Extinction of the Passenger Pigeon." In *Current Ornithology*, 1-33. Vol. 9. New York: Plenum Press, 1992.

Buck, Caitlin E., and Edouard Bard. "A Calendar Chronology for Pleistocene Mammoth and Horse Extinction in North America Based on Bayesian Radiocarbon Calibration." *Quaternary Science Reviews* 26, no. 17-18 (September 2007): 2031–35.

Donlan, Josh. "Re-wilding North America." *Nature* 436, no. 7053 (August 18, 2005): 913–14.

Kay, Charles, and Randy T. Simmons. *Wilderness and Political Ecology: Aboriginal Influences and the Original State of Nature.* Salt Lake City: University of Utah Press, 2002.

Martin, Paul S., and Christine R. Szuter. "War Zones and Game Sinks in Lewis and Clark's West." *Conservation Biology* 13, no. 1 (February 1999): 36–45.

Rick, Torben C., and Jon Erlandson. *Human Impacts on Ancient Marine Ecosystems: A Global Perspective.* Berkeley: University of California Press, 2008.

Soulé, Michael, and Reed Noss. "Rewilding and Conservation: Complementary Goals for Continental Conservation." *Wild Earth* 8 (1998): 18–28.

Vuure, T. Van. *Retracing the Aurochs: History, Morphology and Ecology of an Extinct Wild Ox.* Sofia: Pensoft, 2005.

Worster, Donald. *Nature's Economy: A History of Ecological Ideas.* Cambridge: Cambridge University Press, 1994.

5. A BEAUTIFUL WORLD

Leopold, Aldo, and Charles Walsh Schwartz. *A Sand County Almanac, and Sketches Here and There*. New York: Oxford University Press, 1987.

Steinbeck, John. *The Log from the Sea of Cortez*. New York: Viking Press, 1941.

Stutchbury, Bridget Joan. *Silence of the Songbirds*. New York: Walker & Co., 2007.

Thomas, Keith. *Man and the Natural World: A History of the Modern Sensibility*. New York: Pantheon Books, 1983.

6. GHOST ACRES

Barlow, Connie C. *The Ghosts of Evolution: Nonsensical Fruit, Missing Partners, and Other Ecological Anachronisms*. New York: Basic Books, 2000.

Byers, John A. *Built for Speed: A Year in the Life of Pronghorn*. Cambridge, MA: Harvard University Press, 2003.

Lost Life: England's Lost and Threatened Species. Natural England, 2010.

Rackham, Oliver. *The History of the Countryside*. London: J.M. Dent, 1986.

Simmons, I. G. *An Environmental History of Great Britain: From 10,000 Years Ago to the Present*. Edinburgh: Edinburgh University Press, 2001.

Stewart, George R. *Names on the Globe*. New York: Oxford University Press, 1975.

7. UNCERTAIN NATURE

Bradshaw, G. A., Allan N. Schore, Janine L. Brown, Joyce H. Poole, and Cynthia J. Moss. "Elephant Breakdown." *Nature* 433, no. 7028 (February 24, 2005): 807.

Dagg, Anne Innis. *The Social Behavior of Older Animals.* Baltimore, MD: Johns Hopkins University Press, 2009.

Lotze, Heike K., Karsten Reise, Boris Worm, Justus Van Beusekom, Mette Busch, Anneli Ehlers, Dirk Heinrich, Richard C. Hoffmann, Poul Holm, Charlotte Jensen, Otto S. Knottnerus, Nicole Langhanki, Wietske Prummel, Manfred Vollmer, and Wim J. Wolff. "Human Transformations of the Wadden Sea Ecosystem through Time: A Synthesis." *Helgoland Marine Research* 59, no. 1 (April 2005).

Pauly, Daniel, Villy Christensen, Johanna Dalsgaard, Rainer Froese, and Francisco Torres, Jr. "Fishing Down Marine Food Webs." *Science* 279, no. 5352 (February 6, 1998): 860–63.

Terborgh, John, and J. A. Estes. *Trophic Cascades: Predators, Prey, and the Changing Dynamics of Nature.* Washington, DC: Island Press, 2010.

8. WHAT NATURE LOOKS LIKE

DeMartini, E. E., A. M. Friedlander, S. A. Sandin, and E. Sala. "Differences in Fish-assemblage Structure between Fished and Unfished Atolls in the Northern Line Islands, Central Pacific." *Marine Ecology Progress Series* 365 (August 18, 2008): 199–215.

Haggan, N., A. Beattie, and D. Pauly, eds. "Back to the Future:

Reconstructing the Hecate Strait Ecosystem." *Fisheries Centre Research Reports* 7, no. 3 (1999): 65.

Pandolfi, J. M. "Global Trajectories of the Long-Term Decline of Coral Reef Ecosystems." *Science* 301, no. 5635 (2003): 955–58.

Smetacek, Victor. "Are Declining Antarctic Krill Stocks a Result of Global Warming or of the Decimation of the Whales?" In *Impacts of Global Warming on Polar Ecosystems*, edited by Carlos M. Duarte, 45–83. Fundación BBVA, 2008.

The IUCN Red List of Threatened Species. www.iucnredlist.org.

9. THE MAKER AND THE MADE

Chivian, Eric, and Aaron Bernstein. *Sustaining Life: How Human Health Depends on Biodiversity*. Oxford: Oxford University Press, 2008.

Crutzen, Paul J. "Geology of Mankind." *Nature* 415 (January 2002): 23.

MacArthur, Robert H., and Edward O. Wilson. *The Theory of Island Biogeography*. Princeton, NJ: Princeton University Press, 1967.

Mora, Camilo, Derek P. Tittensor, Sina Adli, Alastair G. B. Simpson, and Boris Worm. "How Many Species Are There on Earth and in the Ocean?" *PLoS Biology* 9, no. 8 (August 2011).

Quammen, David. *The Song of the Dodo: Island Biogeography in an Age of Extinctions*. New York: Scribner, 1996.

Wade, Nicholas. *Before the Dawn: Recovering the Lost History of Our Ancestors*. New York: Penguin Press, 2006.

Wilson, Edward O. *The Diversity of Life*. Cambridge, MA: Belknap Press of Harvard University Press, 1992.

10. THE AGE OF REWILDING

Campbell, Karl, and C. Josh Donlan. "Feral Goat Eradications on Islands." *Conservation Biology* 19, no. 5 (October 2005): 1362–74.

Cronon, William, ed. *Uncommon Ground: Toward Reinventing Nature.* New York: W.W. Norton & Company, 1995.

Jackson, S. T., and R. J. Hobbs. "Ecological Restoration in the Light of Ecological History." *Science* 325, no. 5940 (July 30, 2009): 567–69.

Soulé, Michael E., and Gary Lease. *Reinventing Nature?: Responses to Postmodern Deconstruction.* Washington, D.C.: Island Press, 1995.

11. DOUBLE DISAPPEARANCE

Kirch, Patrick V. "Hawaii as a Model System for Human Ecodynamics." *American Anthropologist* 109, no. 1 (March 2007): 8–26.

Kittinger, John N., John M. Pandolfi, Jonathan H. Blodgett, Terry L. Hunt, Hong Jiang, Kepā Maly, Loren E. McClenachan, Jennifer K. Schultz, and Bruce A. Wilcox. "Historical Reconstruction Reveals Recovery in Hawaiian Coral Reefs." Edited by Stuart A. Sandin. *PLoS ONE* 6, no. 10 (October 3, 2011).

Limburg, Karin E., and John R. Waldman. "Dramatic Declines in North Atlantic Diadromous Fishes." *BioScience* 59, no. 11 (December 2009): 955–65.

Rogers, Raymond Albert. *Nature and the Crisis of Modernity: A Critique of Contemporary Discourse on Managing the Earth.* Montréal: Black Rose Books, 1994.

Steadman, D. W. "Prehistoric Extinctions of Pacific Island Birds: Biodiversity Meets Zooarchaeology." *Science* 267, no. 5201 (February 24, 1995): 1123–31.

Vitousek, P. M. "Soils, Agriculture, and Society in Precontact Hawai'i." *Science* 304, no. 5677 (June 11, 2004).

12. THE LOST ISLAND

Hunt, Terry L., and Carl P. Lipo. *The Statues That Walked: Unraveling the Mystery of Easter Island*. New York: Free Press, 2011.

Quammen, David. "Planet of Weeds: Tallying the Losses of Earth's Animals and Plants." *Harper's*, October 1998, 57–69.

Wilson, Edward O. *The Future of Life*. New York: Alfred A. Knopf, 2002.

EPILOGUE

Gailus, Jeff, and Phil Condon. *Original Griz: The History and Future of the Great Plains Grizzly*. Diss., University of Montana, 2007.

Glavin, Terry. *Waiting for the Macaws: And Other Stories from the Age of Extinctions*. Toronto: Viking Canada, 2006.

Livingston, John A. *Rogue Primate: An Exploration of Human Domestication*. Toronto: Key Porter Books, 1994.

Acknowledgements

More people helped with or inspired aspects of this book than I can hope to thank in this space; most of their names appear in these pages or among the citations. If you happen to be one of them, please know that your contribution was deeply appreciated. Several people made efforts on my behalf that were truly decisive. Jennifer Jacquet frequently broke trail for me into the world of historical ecology, and once even gave me a bed to sleep in on the road. Some of this book's material first appeared in articles for the magazines *Explore*, *Orion* and *The Walrus*, and also as a chapter in *Applying Marine Historical Ecology to Conservation and Management* (University of California Press, 2013); I'd like to acknowledge my respective editors James Little, Andrew Blechman and Jeremy Keehn, and for the U of C Press project, Jack Kittinger, Loren McClenachan, Keryn Gedan and Louise Blight. Anne McDermid, as always, believed in this book from the first. My editors Anne Collins and Courtney Young inspired critical improvements. As always, Alisa Smith walked with me every step of a journey that passed at times through the shadows. I really never can thank her enough.

Index

J.B. MACKINNON has won numerous national and international awards for journalism. As the originator of the 100-mile diet concept, he appears regularly in Canada and the U.S. as a speaker and commentator on the ecology of food. His book *The 100-Mile Diet*, co-authored with Alisa Smith, was a national bestseller and inspired a TV series in which the small town of Mission, B.C., learned to eat locally. He was also the co-author, with Mia Kirshner and artists Paul Shoebridge and Michael Simons, of *I Live Here*, a groundbreaking "paper documentary" about displaced people that made top ten lists in media as diverse as the *Bloomsbury Literary Review* and *Comic Book Resources*, as well as becoming a *Los Angeles Times* bestseller. His first book, *Dead Man in Paradise*, in which he investigated the assassination of his uncle, a radical priest in the Dominican Republic, won the Charles Taylor Prize for Literary Non-Fiction.